Ergonomics for Beginners

A Quick Reference Guide

Third Edition

Jan Dul • Bernard Weerdmeester

Ergonomics for Beginners

A Quick Reference Guide

Third Edition

CRC Press
Taylor & Francis Group
Boca Raton London New York

CRC Press is an imprint of the
Taylor & Francis Group, an **informa** business

CRC Press
Taylor & Francis Group
6000 Broken Sound Parkway NW, Suite 300
Boca Raton, FL 33487-2742

© 2008 by Jan Dul and Bernhard Weerdmeester
CRC Press is an imprint of Taylor & Francis Group, an Informa business

No claim to original U.S. Government works
Printed in the United States of America on acid-free paper
10 9 8 7 6 5 4 3

International Standard Book Number-13: 978-1-4200-7751-3 (Softcover)

Library of Congress Cataloging-in-Publication Data

Dul, Jan, 1954-
    Ergonomics for beginners : a quick reference guide / Jan Dul, Bernard Weerdmeester. -- 3rd ed.
      p. ; cm.
    Includes bibliographical references and index.
    ISBN 978-1-4200-7751-3 (hardback : alk. paper)
    1. Human engineering. I. Weerdmeester, B. A. (Bernard A.), 1955- II. Title.
    [DNLM: 1. Human Engineering. TA 166 D878e 2008]

    TA166.D78 2008
    620.8'2--dc22   2008012304

Visit the Taylor & Francis Web site at
http://www.taylorandfrancis.com

and the CRC Press Web site at
http://www.crcpress.com

Contents

Foreword

Publication of this third edition of *Ergonomics for Beginners* coincides with the celebration of the 50th birthday of the International Ergonomics Association. Over the last 50 years the field of human factors/ergonomics (HF/E) has matured as a unique and independent discipline that focuses on the nature of human–artifact interactions, viewed from the unified perspective of science, engineering, design, technology and management of human-compatible systems. Furthermore, applications of HF/E knowledge over the last few decades facilitated significant improvements in usability, safety, performance, and effectiveness and reliability of a variety of work systems, leading, for example, to the reduction of work-related healthcare costs and improving the quality of processes and products, as the well as quality of working life for millions of people worldwide.

As the ergonomics profession promotes a holistic approach to system design by taking into consideration a broad set of interacting physical, cognitive, social, organizational, environmental and other relevant factors, the current book should help enhance the public understanding of the inherent value of the human-centered approach to the design of work systems, workplaces or workstations.

This unique and very important book should find its way into the hands of many individuals who aim to educate themselves about the basic ergonomics principles relevant to the design and use of products, services and systems. The book should also help readers improve their understanding of both the utility and limitations of modern technology. The design of continuously evolving human interactions with technology and work systems requires (in general) an involvement of ergonomically competent persons. HF/E-literate persons, including students and professionals from many different fields and indeed citizens at large, should know enough about how technological systems operate that they are able to make informed choices and benefit from what those technological products and systems afford.

The requisite ergonomics literacy prepares individuals to perform their roles in the workplace and outside the working environment. An ergonomically literate person can learn enough about how technological systems operate to protect her/himself by making informed choices and making use of beneficial affordances of the artifacts and environment. The standards for HF/E literacy should involve the requisite HF/E knowledge and skills, ways of thinking and acting and practical HF/E capabilities. In this context, people should have a basic knowledge of the philosophy of human-centered design and the principles for accommodating human limitations. They should also seek information about the benefits and risks of consumer products, services, and systems and use this knowledge in making decisions about purchasing and using those products and systems. Finally, people should be able to identify and solve simple task-related design problems at work or home and to apply basic HF/E concepts of human-centered design in making informed judgments about the usability of a product and the related risks and benefits of its use.

In view of the above discussion, publication of this third edition of this book should also help meet the rapidly growing need for ergonomics literacy for all people around the world. The elemental, but comprehensive treatment of the subject matter allowed the authors to develop an authoritative and archival reference book of basic theoretical and practical knowledge, design data and other human-centered information that can be universally studied and applied by people inside and outside the HF/E profession.

This book should also help reduce the undesirable effects and unintended consequences of many human interactions with technology, including work-related physical stresses, employee injury and disability, the risks to patients' well-being, or the potential loss of life while in hospital care, or due to industrial accidents, airplane crashes, train derailments or other urban infrastructure–related disturbances. In the long term, adoption of such an approach to a variety of social settings and environments should also help in economic development that would benefit the global society at large.

This book, in a clear and eloquent way, promotes the application of the human-centered approach to the design, testing and evaluation of work systems by considering the interrelated set of physical, cognitive, social, organizational and other relevant human factors. The authors should be congratulated for updating the content of this important book and for making it available to the HF/E community and the public at large at this opportune time.

Waldemar Karwowski
Professor and Chair
Department of Industrial Engineering
and Management Systems
University of Central Florida
Orlando, Florida, USA

Preface

Ergonomics for Beginners is the English edition of the Dutch classic *Vademecum Ergonomie*, which was published for the first time in 1963 and which has been translated into many languages. The first English edition appeared in 1991 and was received very well as a practical introduction to the broad field of ergonomics. Now, in this third edition of *Ergonomics for Beginners* we have made several changes and additions.

In Chapter 2, "Posture and Movement," we have actualized and extended the anthropometrics tables for the design of products and workplaces by including more body dimensions, by adding data for the U.S. population in addition to the British data and by giving estimation techniques for combining body dimensions and combining populations.

Because of the rapid developments in information and communication technology we have again completely revised Chapter 3, "Information and Operation." We now have also included ergonomics advice regarding topics such as wireless, remote and hands-free controls; website design; mobile interaction; and virtual reality.

To better serve our readers, we have extended Chapter 7, "Sources of Additional Information." We have not only updated the list of ergonomics books, journals and websites but also included a list of 171 published ISO standards on a variety of ergonomics topics.

We hope that these and other changes will further strengthen the impact of our book, and will provide you with the introduction to ergonomics that you are looking for.

Jan Dul and Bernard Weerdmeester

1 Introduction

Ergonomics developed into a recognized field during World War II, when for the first time, technology and the human sciences were systematically applied in a coordinated manner. Physiologists, psychologists, anthropologists, medical doctors, work scientists and engineers together addressed the problems arising from the operation of complex military equipment. The results of this interdisciplinary approach appeared so promising that the cooperation was pursued after the war, in industry. Interest in the approach grew rapidly, especially in Europe and the U.S., leading to the foundation in England of the first ever national ergonomics society in 1949, which is when the term *ergonomics* was adopted. This was followed in 1961 by the creation of the International Ergonomics Association (IEA), which represents ergonomics societies that are active in more than 40 countries or regions, with a total membership of some 19,000 people.

1.1 WHAT IS ERGONOMICS?

The word *ergonomics* is derived from the Greek words *ergon* (work) and *nomos* (law). In several countries, the term *human factors* is also used. A succinct definition would be that ergonomics aims to design appliances, technical systems and tasks in such a way as to improve human safety, health, comfort and performance. The formal definition of *ergonomics*, approved by the IEA, reads as follows: Ergonomics (or human factors) is the scientific discipline concerned with understanding of the interactions among humans and other elements of a system, and the profession that applies theory, principles, data and methods to design, in order to optimize human well-being and overall system performance.

In the design of work and everyday-life situations, the focus of ergonomics is man. Unsafe, unhealthy, uncomfortable or inefficient situations at work or in everyday life are avoided by taking account of the physical and psychological capabilities and limitations of humans. A large number of factors play a role in ergonomics; these include body posture and movement (sitting, standing, lifting, pulling and pushing), environmental factors (noise, vibration, illumination, climate, chemical substances), information and operation (information gained visually or through other senses, controls, relation between displays and control), as well as work organization (appropriate tasks, interesting jobs). These factors determine to a large extent safety, health, comfort and efficient performance at work and in everyday life. Ergonomics draws knowledge from various fields in the human sciences and technology, including anthropometrics, biomechanics, physiology, psychology, toxicology, mechanical engineering, industrial design, information technology and management. It has gathered selected and integrated relevant knowledge from these fields. In applying this knowledge, specific methods and techniques are used. Ergonomics differs from other fields by its interdisciplinary approach and applied nature. The interdisciplinary

1

character of the ergonomic approach means that it relates to many different human facets. As a consequence of its applied nature, the ergonomic approach results in the adaptation of the workplace or environment to fit people, rather than the other way round.

1.2   WHAT IS AN ERGONOMIST?

In some countries it is possible to graduate as an ergonomist. Other people who are trained in one of the relevant basic technical, medical or social science fields can also acquire knowledge of, and capabilities in, ergonomics through training and experience. In several countries, independent certifying bodies can certify professional ergonomists. In Europe, for example the Center for Registration of European Ergonomists (CREE) decides on candidates for registration as European Ergonomists (EurErg). Similarly in the USA, the Board of Certification in Professional Ergonomics (BCPE) awards the title Certified Professional Ergonomist (CPE). Professional ergonomists can work for the authorities (legislation), training institutions (universities and colleges), research establishments, the service industry (consultancy) and the production sector (occupational health services, personnel departments, design departments, research departments, etc.).

Many professional ergonomists who are active in business (company ergonomists) practice their profession mainly by being an intermediary between, on the one hand, the designers and, on the other hand, the users of production systems. The ergonomist highlights the areas where ergonomic knowledge is essential, provides ergonomic guidelines and advises designers, purchasers, management and employees on which are the more acceptable systems. There are other experts besides professional ergonomists who make use of ergonomic knowledge, methods and techniques. These would include, for example, industrial engineers, industrial designers, company doctors, company nurses, physiotherapists, industrial hygienists and industrial psychologists.

1.3   THE SOCIAL VALUE OF ERGONOMICS

Ergonomics can contribute to human well-being in terms of safety, health, and comfort. Daily occurrences such as accidents at work, in traffic and at home, as well as disasters involving cranes, airplanes and nuclear power stations can often be attributed to human error. From the analysis of these failures it appears that the cause is often a poor and inadequate relationship between operators and their task. The probability of accidents can be reduced by taking better account of human capabilities and limitations when designing work and everyday-life environments. In the design of complex technical systems such as process installations, aircraft and (nuclear) power stations, ergonomics has become one of the most important design factors in reducing operator error.

Many work and everyday-life situations are hazardous to health. In many countries, diseases of the musculoskeletal system (mainly lower back pain) and psychological illnesses (for example, due to stress) constitute the most important causes of absence due to illness, and of occupational disability. These conditions can be partly ascribed to poor design of equipment, technical systems and tasks. Here, too, ergonomics can help reduce the problems by improving the working conditions. Therefore, in a number of countries, occupational health and safety regulations refer to ergonomics as a means to prevent work-related health problems. Some ergonomic knowledge has been compiled into official standards whose objectives are to stimulate the application of ergonomics and to prevent health problems. A range of ergonomic subjects is covered by international ISO standards of the International Standardization Organization (ISO), European EN-standards of the Comité European de Normalisation (CEN), as well as national standards. In addition, there are specific ergonomic standards, which are applied in individual companies and in industrial sectors.

Finally, ergonomics can contribute to the realization of user-friendly products. For example, many consumer products (e.g., input devices for computers) are being promoted as ergonomical, suggesting comfort and pleasure during the use of the product.

1.4   THE ECONOMIC VALUE OF ERGONOMICS

By definition, ergonomics can serve both social goals (well-being) and economic goals (performance). At society level, ergonomics can contribute to the reduction of costs due to preventable health problems such as work-related musculoskeletal disorders by improving working conditions. The societal costs include health care costs for the treatment of disorders and costs related to the loss of labor productivity due to absence from work.

At company level, ergonomics can contribute to the competitive advantage of a company. With ergonomically designed production processes, a company can increase human performance in terms of productivity and quality, and can realize important cost-reductions. Furthermore, with ergonomically designed products, a company can deliver benefits to its customers, which exceed those of competing products.

1.5   GENERAL AND INDIVIDUAL ERGONOMICS

An important ergonomic principle is that equipment, technical systems and tasks have to be designed in such a way that they are suited to every user. The variability within populations is such that most designs, in the first instance, are suited to only 95 per cent of the population. This means that the design is less than optimum for five per cent of the users, who then require special, individual ergonomic measures. Examples of groups of users, who from an ergonomic perspective require additional attention, are short or tall persons, overweight people, the handicapped, the old, the young and pregnant women.

This book focuses primarily on the application of ergonomics in a more general sense. Individual requirements for special groups cannot be dealt with in a book of this size.

2 Posture and Movement

Posture and movement play a central role in ergonomics. At work and in everyday life, postures and movements are often imposed by the task and the workplace. The body's muscles, ligaments and joints are involved in adopting a posture, carrying out a movement and applying a force. The muscles provide the force necessary to adopt a posture or make a movement. The ligaments, on the other hand, have an auxiliary function, while the joints allow the relative movement of the various parts of the body. Poor posture and movement can lead to local mechanical stress on the muscles, ligaments and joints, resulting in complaints of the neck, back, shoulder, wrist and other parts of the musculoskeletal system. Some movements not only produce a local mechanical stress on the muscles and joints but also require an expenditure of energy on the part of the muscles, heart and lungs. In the following sections we shall begin by providing some general background on posture and movement. Thereafter, possibilities for optimizing tasks and the workplace are presented for commonplace postures and movements such as sitting, standing, lifting, pulling and pushing.

2.1 BIOMECHANICAL, PHYSIOLOGICAL AND ANTHROPOMETRIC BACKGROUND

A number of principles of importance to the ergonomics of posture and movement derive from a range of specialist fields, namely biomechanics, physiology and anthropometrics. These general principles are discussed in this section and are applied in the subsequent sections (see pp. 12 and 28) to some specific postures and movements.

2.1.1 BIOMECHANICAL BACKGROUND

In biomechanics, the physical laws of mechanics are applied to the human body. It is thereby possible to estimate the local mechanical stress on muscles and joints, which occurs while adopting a posture or making a movement. A few biomechanical principles of importance to the ergonomics of posture and movement are outlined below.

Joints must be in a neutral position

When maintaining a posture or making a movement, the joints ought to be kept as much as possible in a neutral position. In the neutral position the muscles and ligaments that span the joints are stretched to the least possible extent and are thus subject to less stress. In addition, the muscles are able to deliver their greatest force when the joints are in the neutral position.

Raised arms, bent wrists, bent neck and turned head, and bent and twisted trunk are examples of poor postures where the joints are not in a neutral position.

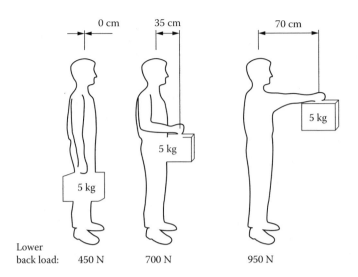

FIGURE 2.1   Increasing the distance between the hands and the body increases the stress on, among other areas, the lower back (10 N is about 1 kg-force).

Keep the work close to the body

If the work is too far from the body, the arms will be outstretched and the trunk bent over forward. The weight of the arms, head, trunk and possibly the weight of any load being held then exerts a greater horizontal leverage on the joints under stress (elbow, shoulder, back). This obviously increases the stress on these muscles and joints. Figure 2.1 shows that the stress to the back increases when the arms are outstretched.

Avoid bending forward

The upper part of the body of an adult weighs about 40 kg on average. The further the trunk is bent forwards, the harder it is for the muscles and ligaments of the back to maintain the upper body in balance. The stress is particularly large in the lower back. Prolonged bending over for long periods must therefore be avoided wherever possible.

A twisted trunk strains the back

Twisted postures of the trunk cause undesirable stress to the spine.

The elastic discs between the vertebrae are stretched, and the joints and muscles on both sides of the spine are subjected to asymmetric stress.

Sudden movements and forces produce peak stresses

Sudden movements and forces can produce large, short-duration stresses. These peak stresses are a consequence of the acceleration in the movement. It is well known that sudden lifting can cause acute back pain in the lower back. Lifting must occur as far as possible in an even and gradual manner. Thorough preparation is necessary before large forces are exerted.

Alternate postures as well as movements

No posture or movements should be maintained for a long period of time. Prolonged postures and repetitive movements are tiring and, in the long run, can lead to injuries to the muscles and joints. Although the ill effects of prolonged postures and repetitive movements can be prevented by alternating tasks, it is best to avoid movements that involve regular lifting or repetitive arm movements. Likewise, standing, sitting and walking should also be alternated, and it should be possible to carry out prolonged tasks either standing or sitting.

Limit the duration of any continuous muscular effort

Continuous stress on certain muscles in the body as a result of a prolonged posture or repetitive movement leads to localized muscle fatigue, a state of muscle discomfort and reduced muscle performance. As a result, the posture or movement cannot be maintained continuously. The greater the muscular effort (exerted force as a percentage of the maximum force), the shorter the time it can be maintained (Figure 2.2).

Most people can maintain a maximum muscular effort for no more than a few seconds and a 50 per cent muscular effort for no more than approximately one minute as this causes muscular exhaustion.

Prevent muscular exhaustion

The muscles will take a fairly long time to recover if they become exhausted which is why exhaustion must be avoided. Figure 2.3 shows an example of recovery curves after a muscle has been partially or totally exhausted from continuous effort.

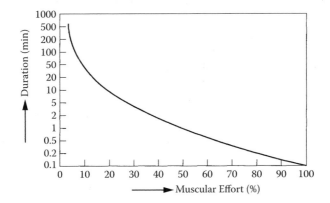

FIGURE 2.2   The duration of continuous localized muscular effort must be limited. The figure shows the relationship between muscular effort (exerted force as a percentage of maximum force) and the maximum possible duration (in minutes) of any continuous muscular effort.

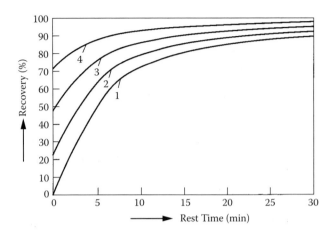

FIGURE 2.3   Recovery curves for muscles which have been exhausted (curve 1) or partially exhausted (curves 2 through 4) after continuous muscular effort.

    In this example, an exhausted muscle needs to rest for 30 minutes to achieve a 90 per cent recovery. Muscles in a half-exhausted state will recover to the same degree after 15 minutes. Complete recovery can take many hours.

More frequent short breaks are better than a single long one

Muscular fatigue can be reduced by distributing the resting time over the task duration or working day. It is not sensible to accumulate break times until the end of the task or working day.

2.1.2   PHYSIOLOGICAL BACKGROUND

In exercise physiology, estimates are made of the energy demands on the heart and lungs resulting from muscular effort during movements. In addition to fatigue that results from continuous localized muscular effort (see Biomechanical Background, p. 5), general body fatigue can develop from carrying out physical tasks over a long period. The limiting factor here is the amount of energy that the heart and lungs can supply to the muscles to allow postures to be adopted or movements to be carried out. A few physiological principles of importance to the ergonomics of posture and movement are discussed below.

Limit the energy expenditure in a task

The majority of the population can carry out a prolonged task without experiencing any general fatigue provided the energy demand of the task (expressed as the energy consumed by the person per unit of time) does not exceed 250 W (1 W = 0.06 kj/min = 0.0143 kcal/min). This figure includes the amount of energy, approximately 80W, which the body needs when at rest. At this energy consumption level, the task is not

TABLE 2.1

Examples of Activities with an Energy Demand in Excess of 250 W

Activity	Energy expenditure
Walking while carrying a load (30 kg, 4 km hr^{-1})	370 W
Frequent lifting (1 kg, 1 × per sec)	600 W
Running (10 km hr^{-1})	670 W
Cycling (20 km hr^{-1})	670 W
Climbing stairs (30 deg, 1 km hr^{-1})	960 W

Note: Additional measures (breaks, alternation with lighter activities, etc.) are necessary to avoid exhaustion in the long term.

considered heavy, and no special measures such as breaks or alternating with light activities are necessary for recovery. Examples of activities with an energy demand of less than 250 W are writing, typing, ironing, assembling light materials, operating machinery, a gentle walk or a leisurely cycle ride.

Rest is necessary after heavy tasks

If the energy demand during a task exceeds 250 W, then additional rest is necessary to recover. Rest can be in the form of breaks or less demanding tasks. The reduction in activity must be such that the average energy demand over the working day does not exceed 250 W.

Table 2.1 lists some activities with a high-energy demand. It is also true here that rest is most effective if the total rest time is spread over a number of break periods spaced regularly during the task and not saved up until the end of the task or the end of the working day.

2.1.3 ANTHROPOMETRIC BACKGROUND

Anthropometrics is concerned with the size and proportions of the human body. A few anthropometric principles of importance to the ergonomics of posture and movements are given below.

Take account of differences in body size

The designers of workplaces, accessories and such must bear in mind differences in body size of the potential users. A table height, which is suitable for a person of average stature, can be unsuitable for a tall or short person. A table height that is adjustable over a sufficient range is the solution if the table is to be used by several people.

Sometimes only the shortest users must be considered, for example, in designing a control panel that has to be reached with the arms. In other cases, such as in choosing a door height, only the tall users have to be considered instead.

Use the anthropometric tables appropriate for specific populations

Data for body dimensions always refer to a particular population group and do not necessarily apply to other population groups. Table 2.2, for example, shows the body dimensions of British and U.S. adults. The adult populations of Great Britain and the U.S. are relatively tall in comparison with the average world population. The dimensions refer to unclothed, unshod persons. Some 30–50 mm must be added to the stature to account for shoe thickness. The data in the table do not apply to other population groups.

TABLE 2.2
Anthropometric Estimates for Adults Aged 19–65 Years

Dimension	BRITISH MEN					BRITISH WOMEN			
	P5	P50	P95	SD		P5	P50	P95	SD
1. Stature	1625	1740	1855	70		1505	1610	1710	62
2. Eye height	1515	1630	1745	69		1405	1505	1610	61
3. Shoulder height	1315	1425	1535	66		1215	1310	1405	58
4. Elbow height	1005	1090	1180	52		930	1005	1085	46
5. Hip height	840	920	1000	50		740	810	885	43
6. Knuckle height	690	755	825	41		660	720	780	36
7. Fingertip height	590	655	720	38		560	625	685	38
8. Sitting height	850	910	965	36		795	850	910	35
9. Sitting eye height	735	790	845	35		685	740	795	33
10. Sitting shoulder height	540	595	645	32		505	555	610	31
11. Sitting elbow height	195	245	295	31		185	235	280	29
12. Thigh thickness	135	160	185	15		125	155	180	17
13. Buttock-knee length	540	595	645	31		520	570	620	30
14. Buttock-popliteal length	440	495	550	32		435	480	530	30
15. Knee height	490	545	595	32		455	500	540	27
16. Popliteal height	395	440	490	29		355	400	445	27
17. Shoulder breadth (bideltoid)	420	465	510	28		355	395	435	24
18. Shoulder breadth (biacromial)	365	400	430	20		325	355	385	18
19. Hip breadth	310	360	405	29		310	370	435	38
20. Chest (bust) depth	215	250	285	22		210	250	295	27
21. Abdominal depth	220	270	325	32		205	255	305	30
22. Shoulder-elbow length	330	365	395	20		300	330	360	17
23. Elbow-fingertip length	440	475	510	21		400	430	460	19
24. Upper limb length	720	780	840	36		655	705	760	32
25. Shoulder-grip length	610	665	715	32		555	600	650	29
26. Head length	180	195	205	8		165	180	190	7
27. Head breadth	145	155	165	6		135	145	150	6
28. Hand length	175	190	205	10		160	175	190	9
29. Hand breadth	80	85	95	5		70	75	85	4
30. Foot length	240	265	285	14		215	235	255	12
31. Foot breadth	85	95	110	6		80	90	100	6
32. Span	1655	1790	1925	83		1490	1605	1725	71
33. Elbow span	865	945	1020	47		780	850	920	43
34. Vertical grip reach (standing)	1925	2060	2190	80		1790	1905	2020	71
35. Vertical grip reach (sitting)	1145	1245	1340	60		1060	1150	1235	53
36. Forward grip reach	720	780	835	34		650	705	755	31
Body mass (kg)	55	75	94	12		44	63	81	11

To gain an idea of the extent of individual variation in body size, the data in Table 2.2 show body dimensions, for men and woman separately, of:

- Short persons (P5: only 5 per cent of persons are shorter)
- The average person (P50)
- Tall persons (P95: only 5 per cent of persons are taller).

For each dimension the standard deviation (SD) is also given as a measure of variation.

TABLE 2.2 (CONTINUED)
Anthropometric Estimates for Adults Aged 19–65 Years

Dimension	U.S. MEN					U.S. WOMEN			
	P5	P50	P95	SD		P5	P50	P95	SD
1. Stature	1640	1755	1870	71		1520	1625	1730	64
2. Eye height	1529	1644	1759	70		1416	1519	1622	63
3. Shoulder height	1330	1440	1550	67		1225	1325	1425	60
4. Elbow height	1020	1105	1190	53		945	1020	1095	47
5. Hip height	835	915	995	50		760	835	910	45
6. Knuckle height	700	765	830	41		670	730	790	37
7. Fingertip height	595	660	725	39		565	630	695	40
8. Sitting height	855	915	975	36		800	860	920	36
9. Sitting eye height	740	800	860	35		690	750	810	35
10. Sitting shoulder height	545	600	655	32		510	565	620	32
11. Sitting elbow height	195	245	295	31		185	235	285	29
12. Thigh thickness	135	160	185	16		125	155	185	17
13. Buttock-knee length	550	600	650	31		525	575	625	31
14. Buttock-popliteal length	445	500	555	33		440	490	540	31
15. Knee height	495	550	605	32		460	505	550	28
16. Popliteal height	395	445	495	29		360	405	450	28
17. Shoulder breadth (bideltoid)	425	470	515	28		360	400	440	25
18. Shoulder breadth (biacromial)	365	400	435	21		330	360	390	19
19. Hip breadth	310	360	410	30		310	375	440	39
20. Chest (bust) depth	220	255	290	22		210	255	300	28
21. Abdominal depth	220	275	330	32		210	260	310	31
22. Shoulder-elbow length	330	365	400	21		305	335	365	18
23. Elbow-fingertip length	445	480	515	21		400	435	470	20
24. Upper limb length	730	790	850	36		655	715	775	35
25. Shoulder-grip length	615	670	725	33		560	610	660	30
26. Head length	180	195	210	8		165	180	195	8
27. Head breadth	145	155	165	6		135	145	155	6
28. Hand length	175	191	205	10		160	175	190	10
29. Hand breadth	80	90	100	5		65	75	85	5
30. Foot length	240	265	290	14		220	240	260	13
31. Foot breadth	90	100	110	6		80	90	100	6
32. Span	1670	1810	1950	84		1505	1625	1745	73
33. Elbow span	875	955	1035	48		790	860	930	44
34. Vertical grip reach (standing)	1950	2080	2210	80		1805	1925	2045	73
35. Vertical grip reach (sitting)	1155	1255	1355	61		1070	1160	1250	55
36. Forward grip reach	725	785	845	35		655	710	765	32
Body mass (kg)	55	78	102	14		41	65	89	15

Source: Data from Pheasant, S. and Haslegrave, C., 2006, *Bodyspace: Anthropometry, Ergonomics and the Design of Work, Third Edition*, CRC Press, Boca Raton, Florida.

TABLE 2.2 (CONTINUED)

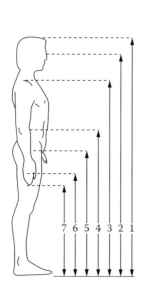

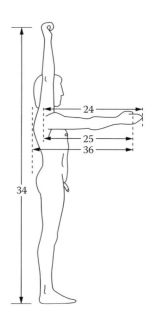

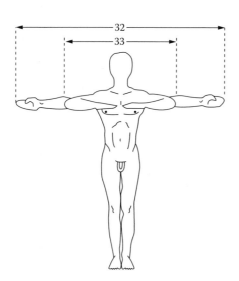

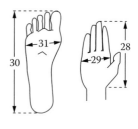

TABLE 2.2 (CONTINUED)

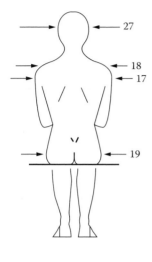

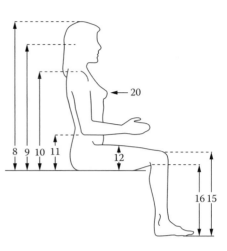

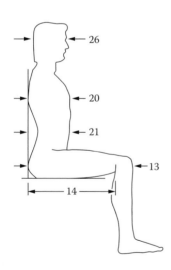

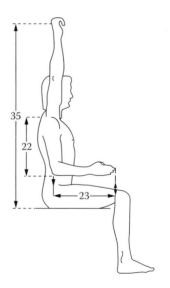

The average height of a British man is 174.0 cm, and of a British woman 161.0 cm; the height of a short British man is 162.5 cm or less, and of a short British woman 150.5 cm or less; and the height of a tall British man is over 185.5 cm, and of a tall British woman is over 171.0 cm. The average height of a U.S. man is 175.5 cm, and of a U.S. woman 162.5 cm; the height of a short U.S. man is 164.0 cm or less, and of a short U.S. woman 152.0 cm or less; and the height of a tall U.S. man is over 187.0 cm, and of a tall U.S. woman over 173.0 cm.

The correlation between body dimensions in Table 2.2 is limited. For instance, a person with a short lower arm (dimension 23 is small) could have a long trunk (dimension 10 is large).

For certain applications in designing products or workplaces, it may be needed to add or subtract dimensions. Under the following conditions two dimensions in Table 2.2. can be added or subtracted:

- The two dimensions must be in line within the studied body posture. For example, the popliteal height (dimension 16) and the sitting elbow height (dimension 11) can be added for getting the sitting elbow height from the floor.
- For adding (+) or subtracting (−) dimensions a and b, the following formulae must be used for calculating for the new dimension (dimension c): the mean ($P50_c$), the standard deviation (SD_c), the 5th percentile ($P5_c$), and the 95th percentile ($P95_c$):

$$P50_c = P50_a \pm P50_b$$

$$SD_c^2 = (SD_a)^2 + (SD_b)^2 \pm 2\,r\,SD_a SD_b$$

$$P5_c = P50_c - 1.65\,SD_c$$

$$P95_c = P50_c + 1.65\,SD_c$$

where r is the correlation between the two dimensions. For two length dimensions and two width dimensions the estimated value of r = 0.65, and for two depth dimensions r = 0.20, for a depth and a length dimension r = 0.20, for a width and a length dimension r = 0.30, and for a depth and a width dimension r = 0.40 (Source: www.DINED.nl).

Length dimensions are 1, 2, 3, 4, 5, 6, 7, 8, 9, 10, 11, 13, 14, 15, 16, 22, 23, 24, 25, 28, 30, 32, 33, 34, 35; width dimensions are 17, 18, 19, 27, 29, 31; and depth dimensions are 12, 20, 21, 26.

For example, when the two length dimensions popliteal height (dimension 16) and sitting elbow height (dimension 11) must be added for getting the sitting elbow height from the floor, for the average British male (P50) this new dimension is 44.0 +

24.5 = 68.5 cm. Because SD for the new dimension of British males is the square root of $(2.9)^2 + (3.1)^2 + 2(0.65)(2.9)(3.1) = 5.45$ cm, the new dimension for short British males is 68.5 − 1.65(5.45) = 59.5 cm. This result is different from just adding dimensions 16 and 11 for short British males (39.5 + 19.5 = 59.0 cm).

The new dimension for tall British males is 68.5 + 1.65(5.45) = 77.5 cm. Adding the dimensions for tall British males would give 49.0 + 29.5 = 78.5.

If two populations are mixed (e.g., males and females, or British males and U.S. males) with the same number of people from each subpopulation, then the body dimensions of the mixed population (m) can be calculated from the body dimensions of the two subpopulations (s1 and s2):

$$P50_m = \tfrac{1}{2}(P50_{s1} + P50_{s2})$$

$$SD_m^2 = \tfrac{1}{2}SD_{s1}^2 + \tfrac{1}{2}SD_{s2}^2 + \tfrac{1}{4}(P50_{s1} - P50_{s2})^2$$

$$P5_m = P50_m - 1.65\,SD_m$$

$$P95_m = P50_m + 1.65\,SD_m$$

It should be noted that a design that is based on estimated anthropometric data must be tested with a sample of users before the design goes to production.

2.2 POSTURE

Posture is often imposed by the task or the workplace. Prolonged postures can in time lead to complaints of the muscles and joints. In this section we take a look at the stress due to prolonged sitting and standing, as well as that due to hand and arm postures, such as would occur in the use of handheld tools.

Select a basic posture that fits the job

The characteristic of the job determines the best basic posture: sitting, standing (p. 24, combinations of sitting and standing (sit–stand work stations p. 21) or work stations with pedestal stools (p. 25). Figure 2.4 provides a selection procedure for the best basic posture.

2.2.1 SITTING

Working for long periods in a seated position occurs mostly in offices, but also occurs in industry (assembly and packaging work, sometimes for machine operation). Sitting has a number of advantages compared to standing. The body is better supported because several support surfaces can be used: floor, seat, backrest, armrest and work surface. Therefore, adopting this posture is less tiring than standing. However, activities that require the operator to exert a lot of force or to move around frequently are best carried out standing (see p. 21).

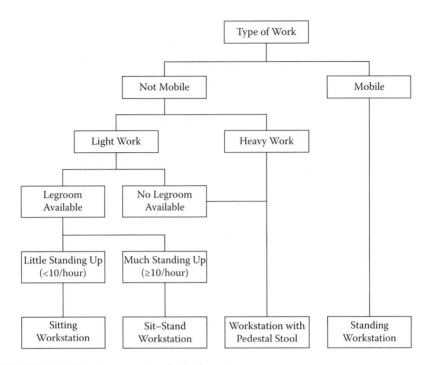

FIGURE 2.4   Selection procedure for basic posture.

Alternate sitting with standing and walking

Although sitting is usually more favorable than standing, sitting for long periods should be avoided as it has a number of disadvantages. Many manual activities carried out while seated (e.g., writing or assembly work) require the person to keep the hands in view. This means that the head and trunk have to be bent forward. The neck and back are then subjected to prolonged stress that can lead to neck and back complaints. Bending the trunk forward also means that the backrest of the chair is no longer used. The back is subject to further stresses if the trunk has to be twisted and the seat cannot swivel. Manual work often requires working with unsupported raised arms, which can lead to shoulder complaints.

Tasks that require prolonged sitting (for example, at a VDU screen) should be alternated with tasks which can be carried out in a standing position, or where walking is necessary. A sit–stand workplace (p. 24) or a chair that promotes active sitting allow the user to alternate between sitting and other postures during the task.

The heights of the seat and backrest of the chair must be adjustable

There are many ergonomically sound chairs on the market. The most important general feature of such chairs is that the height of the seat and backrest is adjustable.

It must be possible while sitting to adjust the height of the seat in a continuous, smooth motion rather than in steps. The height of the seat must be chosen in such a way that when the feet are properly supported, the upper legs are also properly supported, without the back of the knee being cramped.

For British adults, the minimum adjustment range should be at least 13.5 cm between the heights of 38.5 and 52.0 cm, based on popliteal height differences between short females and tall males (measure 16 in Table 2.2 plus 3 cm for shoe thickness). For U.S. adults, the corresponding minimum adjustment range should be at least 13.5 cm between the heights of 39.0 and 52.5 cm.

The backrest must provide support mainly to the lower back; for British and U.S. adults, the recommended adjustment range is 15 cm between the heights of 15 and 30 cm above the seat height, based on differences in the lumbar height, not shown in Table 2.2. Avoid misusing a low backrest as a high backrest.

The lower part of the backrest must be given a convex shape in order to preserve the curve of the lower back.

In addition, the chair should swivel. This reduces the need to twist the body.

Limit the number of adjustment possibilities

Adjustment possibilities must be restricted to only the most important components of the chair, as a minimum, the seat height and the height of the backrest. If too many features are adjustable, settings will be used either incorrectly or not at all.

Provide proper seating instructions

Users of adjustable chairs must receive regular instruction in the optimum adjustment of the chair, say, every six months. This also applies to other adjustable elements of the workplace, such as the table.

Specific chair characteristics are determined by the task

In addition to its general characteristics, an ergonomically sound chair will also display specific features that depend on the task. A chair with armrests can be selected if these do not hinder the activities, as armrests can partly support the weight of the arms and trunk and are also useful when rising from the chair. Armrests should be short to allow close proximity to the table. Castors can be useful if a chair has to be moved frequently but none should be present if pedals have to be operated. If the trunk is mostly upright or tilted somewhat backward, the seat ought to be tilted backward a few degrees. For tasks where the body is unavoidably bent forward, a limited forward tilt (maximum 20 degrees) is advantageous, as it can prevent the lower back from curving. Figure 2.5 shows an example of an ergonomically sound chair for VDU work. Here, the seat and backrest heights are adjustable, the backrest supports also the lower back, and the short armrests and castors provide additional comfort.

The work height depends on the task

The chair is only one of several factors determining whether the working posture is correct. The position of the hands as well as the focal point is also of great importance to the posture of the head, trunk and arms. The correct height for the hands and focal point depends on the task, individual body dimensions and individual prefer-

FIGURE 2.5   An ergonomically sound chair for VDU work. The height of the seat and back rest (with support for the lower back) can easily be adjusted. The chair swivels, has short, adjustable armrests and fitted with castors.

ence. During most tasks the hands have to be used and viewed simultaneously. Then, the work height is a compromise between the optimum height for the arms and the optimum position of the head and trunk. In the first instance, a low table is better since the arms have to be raised to a lesser extent and it is easier to apply a force. In the second instance, a high table is better because it means less bending forward and a better view of the work.

General guidelines for the work height are given in Table 2.3 for three types of tasks. These guidelines apply when both sitting and standing.

TABLE 2.3
Guidelines for the Height of the Hands and Focal Point, for Carrying Out Various Tasks While Seated or Standing

Type of Task	Work Height
Use of eyes: frequent Use of hands/arms: infrequent	10–30 cm below eye height
Use of eyes: frequent Use of hands/arms: frequent	0–15 cm above elbow height
Use of eyes: infrequent Use of hands/arms: frequent	0–30 cm below elbow height

The heights of the hands and focal point need not always be the same as the table or work surface height. The work surface might have to be lowered to take into account the thickness of the work pieces, tools or accessories (e.g., a keyboard). The work surface and the objects on it should not be too thick, otherwise legroom will be restricted.

Work tables used for a given type of seated task not involving objects of different thickness must be adjustable over a range of at least 25 cm because of differences between individuals. Where a number of tasks have to be carried out which require different work heights, the adjustment range must be even greater.

A good starting position for the height of a VDU workstation is one where the hands are kept at elbow height. The height of a VDU table with a keyboard thickness of 3 cm (measured at the position of the middle row of keys) must be adjustable between 51.5 and 74.5 cm for British adults, and between 51.5 and 75 for U.S. adults (based on differences in sitting elbow height between short females and tall males, estimated from dimensions 11 and 16 in Table 2.2, minus 3 cm for keyboard thickness). It must be possible to make the adjustment easily from the seated position.

The heights of the work surface, seat and feet must be compatible

In a seated workplace, which can be adjusted individually, the vertical distances between the feet, seat and working surface must be compatible. The height of the feet is mostly fixed because they rest on the floor. The chair and table must then be adjustable and set according to the guidelines of Table 2.3.

Use a footrest if the work height is fixed

If the individual user cannot adjust the work height, such as at a machine, a relatively high work surface must be chosen to suit tall users. The seat height is then adjusted to the work surface. The height of the feet should then also be adjusted, using a suitable footrest, which would be not simply a bar, but a slightly sloping surface.

Avoid excessive reaches

It is necessary to limit the extent of forward and sideways reaches to avoid having to bend over or twist the trunk. Work pieces, tools and controls that are in regular use should be located directly in front of or near the body.

Figure 2.6 shows the reach envelopes in three planes. The most important operations should take place within a radius of approximately 50 cm. This value applies to both seated and standing work. Application of the guidelines on reaches is given in Chapter 3, which covers the design of control panels.

Select a sloping work surface for reading tasks

If the activities allow it, use should be made of a sloping work surface for reading tasks and other tasks where the work has to be kept in view, such as writing and assembly work without tools. A sloping work surface brings the work to the eye instead of the other way round, thereby improving the posture of the head and trunk.

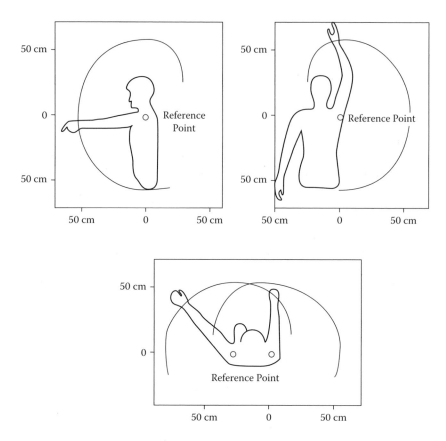

FIGURE 2.6 Guidelines for convenient maximum reaches in seated or standing work.

Because the height of the front of the table or machine remains the same, the arms do not have to be raised any further. Successful use of a sloping work surface requires measures to prevent work pieces or accessories from sliding off (nonslip work surface, rim, etc.), or alternatively, only part of the work surface should slope. Sloping work surfaces can often be created easily by raising the back of the table or machine or by using a lectern. For reading purposes, the position of the work surface that is viewed, must be tilted by at least 45 degrees (Figure 2.7).

For tasks where the hands have to be used and kept in view, such as in writing, the work surface must be placed at an angle of approximately 15 degrees (Figure 2.8). A greater slope is not desirable because of insufficient support for the arms and because objects may slide off.

Allow sufficient legroom

Sufficient legroom must be provided under the work surface (Figure 2.9). The width clearance must be at least 60 cm. The required depth clearance must be at least 40 cm at the knees and 100 cm at the feet, which should allow the user to sit close to the work without bending the trunk forward. It is desirable to be able to stretch the legs

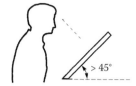

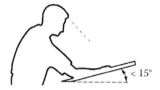

FIGURE 2.7   In tasks that require no manual work, such as reading, bending the head and trunk forward can be reduced by using a sloping work surface of at least 45 degrees for viewing.

FIGURE 2.8   Viewing manual work.

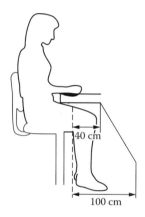

FIGURE 2.9   Required legroom for sitting.

once in a while when sitting for long periods. To this effect the depth clearance at the feet should be at least 1 m. To have sufficient room between the underside of the working surface and the top of the legs, the thickness of the working surface (and objects on it) must be as small as possible. The thickness of a writing surface, for example, should not exceed 3 cm.

2.2.2   STANDING

Activities where considerable force has to be exerted or where the workplace has to be frequently changed should be carried out in a standing position.

Alternate standing with sitting and walking

It is not recommended for the whole working day to be spent in a standing position. Standing for long periods tires the back and legs. An additional stress can arise when the head and trunk are bent, leading to neck and back complaints. Furthermore, working with the arms unsupported, in a raised position, leads to stress on the shoulders, which may result in shoulder complaints. Tasks which have to be carried out over long periods in a standing position should be alternated with tasks which can be carried out while seated or with tasks where walking is required. People should also be given the opportunity to sit down during natural breaks in the work (e.g., in the case of operating a machine or in sales work in shops). A sit–stand workplace or a pedestal stool will allow the user to vary postures during the task (see pp. 24 and 25).

The work height depends on the task

The work height for standing work depends, as for seated work, on the task, on individual preference, and on individual body dimensions. Table 2.3 contains the guidelines for the optimum work height for different types of standing tasks carried out at a work surface.

The height of the work table must be adjustable

It must be possible to adjust the height of a work table that is intended for use by several people (as a result of part-time working, team work or task rotation), or whenever different tasks (e.g., with varying sizes of work pieces) must be carried out at the same table. It must be possible to conveniently adjust the table from the normal working position. A table meant for standing work which is used for a given task, and on which no objects of different thickness are used, must have an adjustment range of at least 25 cm in order to cater to individual differences in body size. Users must be instructed in the optimum height of the table.

Do not use platforms

The use of platforms for standing work is not advisable. The major disadvantages of platforms are that they constitute a trip hazard, are cumbersome to clean, and hamper transport along floors. They also require additional work space and are not practical if their height has to be regularly adjusted for different people or to different working heights.

Provide sufficient room for the legs and feet

Sufficient room must also be kept free under the work surface or machine for the legs and feet in standing work. This allows the person to be close to the work without bending the trunk. Enough clearance is also required for changing the position of the legs once in a while. Figure 2.10 illustrates the required minimum recesses under the work surface or machine.

Avoid excessive reaches

Forward and sideways reaches must be limited in order to avoid having to bend forward or twist the trunk. Work pieces, tools and controls, which are in regular

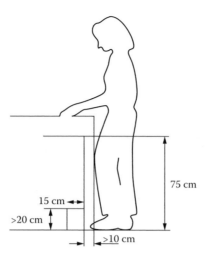

FIGURE 2.10 Minimum foot and legroom required for standing work.

use, should be located directly in front of and near the body. Convenient maximum reaches are given in Figure 2.6.

Select a sloping work surface for reading tasks

If the activities allow it, a sloping work surface should be used for reading tasks just as in the case of seated work. The same is true also for other tasks where the work must be kept in view, such as writing. Guidelines for a sloping work surface are given in Figures 2.7 and 2.8.

2.2.3 CHANGE OF POSTURE

This section describes ways of relieving prolonged postures. These techniques relate to the provision of a varied task package, the implementation of a sit–stand work-place and the use of a pedestal stool.

Offer variation in tasks and activities

The design and organization of activities should ensure that everyone is given varia-tion in tasks and activities so that no prolonged postures occur. The principle of job enrichment can be usefully applied (see Chapter 5).

Introduce sit–stand work stations

If tasks have to be carried out over a long period, the workplace should be adapted to allow the work to be carried out either standing or sitting. To this effect a work height is selected which is suitable for standing work (see Standing, p. 21). In addition, a special high chair also allows the work to be carried out while seated. Legroom should be left under the work surface and a footrest provided (Figure 2.11).

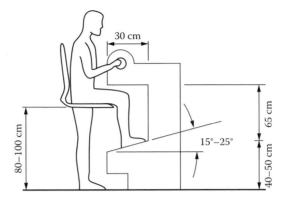

FIGURE 2.11 Guidelines for the dimensions of a workplace at which seated and standing work can be alternated.

Alternate sitting postures

A prolonged seated posture can be varied by using different types of chairs. There are chairs available that promote "active sitting." The chairs offer possibilities to change posture and have adjustable seats and backs, allowing movements of the body. Despite the use of these chairs, it is still advisable to alternate sitting with standing and walking.

Make occasional use of a pedestal stool in standing work

A pedestal stool can be used once in a while to vary a standing work posture. A pedestal stool consists of a seat, which is adjustable in height (65–85 cm), and is tilted forward between 15 and 30 degrees. It allows semi-supported postures to be adopted, which somewhat relieve the stress on the legs. A pedestal stool cannot be used for long periods and is only suited to standing activities where large forces or extensive movements are not required. The floor on which it rests must provide sufficient friction to prevent the support from sliding away (Figure 2.12).

2.2.4 HAND AND ARM POSTURES

Working for long periods with the hand and arm in a poor posture can lead to specific complaints of the wrist, elbow and shoulder. A continuously bent wrist can lead to local nerves becoming inflamed and trapped, resulting in wrist pain and a tingling sensation in the fingers. Another ailment is tennis elbow, which is a local inflammation of a tendon attachment due to a combination of a bent elbow and bent wrist.

Neck and shoulder complaints occur in prolonged work with unsupported, raised arms. These problems arise especially from handling tools. In addition to posture, application of a force and repetitive movement (Repetitive Strain Injuries or RSI) plays a role in the development and aggravation of these conditions. Correct hand and arm postures can be promoted by selecting the correct working height for the hands (see Table 2.3) and by using the right tools (see below).

FIGURE 2.12   A pedestal stool can be used to change posture when standing for long periods.

Select the right tool model

A particular tool is often available in different models. Select a model that is best suited to the task and posture, so that the joints can be kept as much as possible in the neutral position. Figure 2.13 shows the correct and incorrect uses of different types of electric drills and screwdrivers.

Do not bend the wrist, use curved tools instead

Bending the wrist can be prevented by correctly locating the handgrips on a tool (Figure 2.14).

Handheld tools must not be too heavy

If the tool cannot rest on a surface and is normally used with one hand, its weight should not exceed 2 kg. If the tool can rest on a surface, heavier weights are allowed, but the maximum weights that apply to lifting (p. 31) must be taken into consideration.

Heavy tools that are frequently used can be suspended on a counterweight (Figure 2.15).

Maintain your tools

Proper maintenance of tools can contribute to a reduction in bodily stress. Blunt knives, saws or other equipment require greater force. Proper maintenance of motorized handheld tools can also reduce wear, noise and vibration (see Chapter 4).

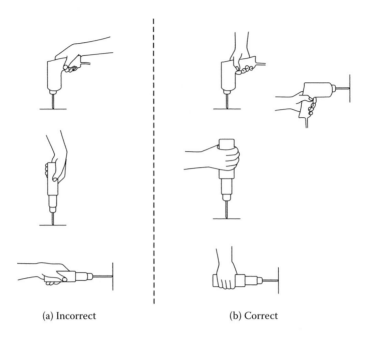

(a) Incorrect   (b) Correct

FIGURE 2.13   When using handheld tools, the wrist should be kept as straight as possible. The figure shows the correct and incorrect use of two types of rotating tool.

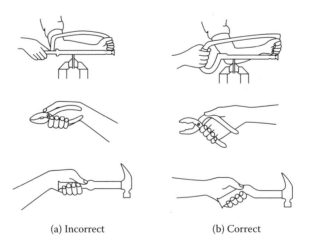

(a) Incorrect   (b) Correct

FIGURE 2.14   Correct location of handgrips on tools avoids having to bend the wrist.

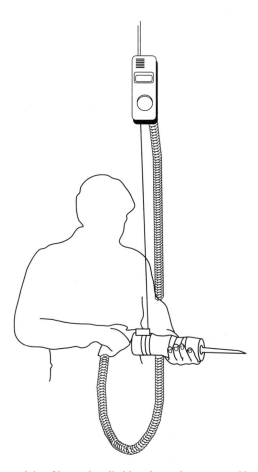

FIGURE 2.15   The weight of heavy handheld tools can be supported by a counterweight.

Pay attention to the shape of handgrips

The shape and location of handgrips on trolleys, loads, machines, equipment and such must take into consideration the position of the hands and arms. If the whole hand is used to exert a force, the handgrip must have a diameter of approximately 3 cm and a length of approximately 10 cm (Figure 2.16). The handgrip must be somewhat convex to increase the contact surface with the hand. The use of preshaped handgrips is not advised: the fingers are constrained, too little account is taken of individual differences in finger thickness and the grips are not suitable for use with gloves.

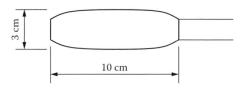

FIGURE 2.16   Shape of handgrips on handheld tools.

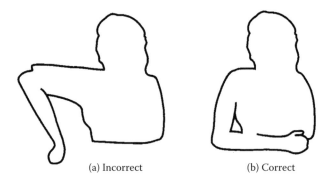

(a) Incorrect   (b) Correct

FIGURE 2.17   Hand and elbow positions above shoulder height are to be avoided.

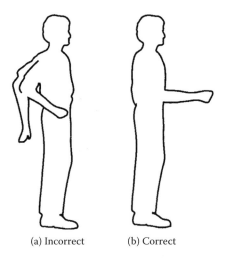

(a) Incorrect   (b) Correct

FIGURE 2.18   Avoid positions where the hands and elbows are held behind the body.

Avoid carrying out tasks above shoulder level

The hands and elbows should be well below shoulder level when carrying out a task
(Figure 2.17). If work above shoulder level is unavoidable, the duration of the work
must be limited and regular breaks must be taken.

Avoid working with the hands behind the body

Working with the hands behind the body should be avoided (Figure 2.18). This
kind of posture occurs when sliding away objects, for example, at check-outs in
supermarkets.

2.3   MOVEMENT

Various tasks require moving the whole body, often while exerting a force. Such movements can cause high, localized mechanical stresses that in time can lead to bodily aches and pains. Movements can also be stressful in the energetic sense for the muscles, heart and lungs. In this section we examine the stress from lifting, carrying, pulling and pushing.

2.3.1   LIFTING

Manual lifting is still frequently needed in spite of mechanization and automation. Lifting can be acceptable if it satisfies certain ergonomic requirements.

This section contains guidelines and measures for acceptable lifting. These measures relate to the production technique (mechanical or manual), the work organization (task design, lifting frequency), the workplace (position of load with respect to the body), the load (shape, weight, presence of handgrips), lifting accessories, and the working method (lifting by several persons, individual lifting technique).

Restrict the number of tasks that require displacing loads manually

Production systems must be designed to use mechanization as a way of restricting the extent of manual lifting. In this case, however, attention must be paid to new problems relating to posture and movement. These could include prolonged manual operation of machines or lifting accessories, or the necessity for heavy maintenance work on machines that are difficult to access. Other problems can also develop as a result of mechanization, for example, noise and vibration, monotony, and reduced social contacts. If it is not possible to avoid heavy or frequent lifting, these activities must be alternated with other (light) activities, for example, by applying job enrichment (see Chapter 5).

In lifting, but also in other physical activities, it is important that the person involved should set the work pace. It is essential to avoid situations where the rate of lifting is imposed by a machine, by colleagues or by a supervisor.

Create optimum circumstances for lifting

If manual lifting of heavy loads (up to 23 kg) is necessary, then lifting conditions have to be optimized:

- It must be possible to hold the load close to the body (horizontal distance from hand to ankles about 25 cm).
- The initial height of the load before it is lifted should be about 75 cm.
- The vertical displacement of the load should not exceed 25 cm.
- It must be possible to pick up the load with both hands.
- The load must be fitted with handles or hand-hold cut-outs.
- It must be possible to choose the lifting posture freely.
- The trunk should not be twisted when lifting.
- The lifting frequency should be less than one lift per five minutes.

- The lifting task should not last more than one hour, and should be followed by a resting time (or light activity) of 120 per cent of the duration of the lifting task.

Ensure that people always lift less, and preferably much less, than 23 kg

Only in the above-mentioned optimum lifting situation may a person lift the maximum load of 23 kg. Lifting conditions, however, are virtually never optimum, in which case the maximum allowable load is considerably less (see below). The load should not exceed a few kilograms if it has to be picked up far away in front of the body and has to be displaced over a large vertical distance.

Use the NIOSH method to assess lifting situations

In practice, optimum lifting conditions are seldom met and, therefore, the permitted load is much less than 23 kg. The method developed by the American National Institute for Occupational Safety and Health (NIOSH) can be used to determine the maximum load in unfavorable lifting conditions. This takes into account the horizontal and vertical distance between the load and the body, the trunk rotation, the vertical displacement, the lifting frequency and duration, and the coupling between hands and load. It assumes among other things that the lifting posture can be freely chosen and that the load is lifted with both hands. The NIOSH equation is devised in such a way that the weight is acceptable for the majority of the population (99 per cent of men and 75 per cent of women), that the compressive load on the lower back is less than 340 ON (340 kg force) and that the energy expenditure for 1–2 hours repetitive lifting is less than 26 OW for lifts below the bench height (75 cm) and less than 19 OW for lifts above the bench height. In the NIOSH method, the unit weight of 23 kg is reduced for unfavorable lifting conditions by using a series of multipliers according to the formula:

Recommended weight limit $= 23$ kg $\times$ HM $\times$ VM $\times$ DM $\times$ FM $\times$ AM $\times$ cm

Figure 2.19 shows the multipliers for the horizontal load distance (horizontal multiplier, HM), the vertical load distance (vertical multiplier, VM), the vertical displacement of the load (displacement multiplier, DM), the frequency (frequency multiplier, FM), the asymmetric factor (asymmetric multiplier, AM), and the coupling factor (coupling multiplier, CM). If the lifting situation does not satisfy the requirements of the NIOSH method (e.g., if the lifting posture cannot be freely chosen, or if the load is lifted with one hand), the method will result in values that are too high. Because of the complexity of the analysis, several software packages have been developed to analyze lifting situations, including combinations of different lifting tasks, using the NIOSH method. These can also help develop improvements based on the results of the analysis.

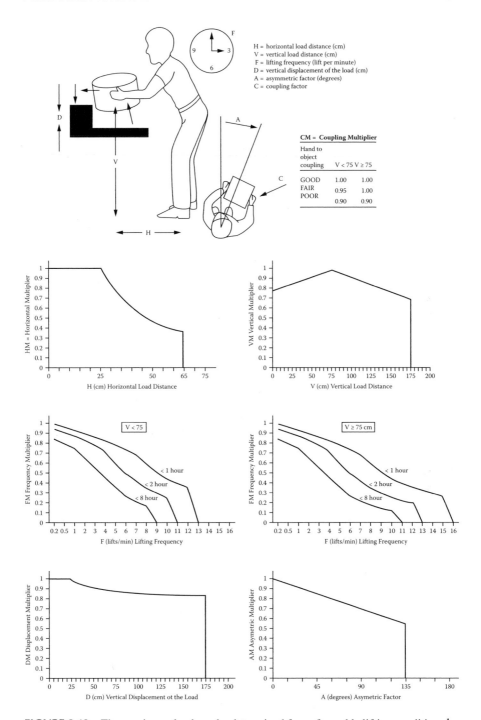

FIGURE 2.19 The maximum load can be determined for unfavorable lifting conditions by using the NIOSH equation.

Individual loads should not be too light

The weight of a load (e.g., the unit weight of a packaging) has to be chosen carefully. On the one hand the NIOSH recommendation should not be exceeded under normal conditions. On the other hand the loads should not be too light otherwise more frequent lifting becomes necessary. If individual loads are too light there is also a danger that several loads may be lifted simultaneously.

Make the workplace suitable for lifting activities

The design of tables, shelves, machines and such, onto which loads have to be placed or from which loads have to be lifted, must result in optimum lifting conditions being achieved.

It must be possible to approach the load properly when lifting and setting it down.

- Foot and legroom must be sufficient to allow a stable position for the feet and to be able to bend the knees.
- Twisting the trunk should not be necessary.
- The height and location of the load on the work surface must be such that when lifting the load or setting it down, the hands are at the optimum height of approximately 75 cm and are close to the trunk.

It can generally be said that the measures described here have considerable influence on the admissibility of lifting activities. It is often possible to achieve more through a higher and closer position of the load than by reducing its weight.

Loads should be fitted with handgrips

A load should be fitted with two handgrips so that it can be grasped with both hands and lifted (Figure 2.20b). Grasping the load with the fingers (Figure 2.20a) should be avoided because far less force can be exerted.

The position of the handgrips should be such that the load cannot twist when lifted.

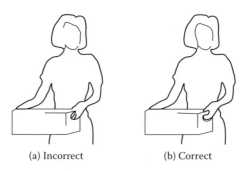

(a) Incorrect   (b) Correct

FIGURE 2.20 No handgrips (a) and correct handgrips (b) for lifting loads.

Ensure that the load is of the correct shape

The size of the load must be as small as possible so that it can be held close to the body. It must be possible to move the load between the knees if it has to be lifted from the floor. The load should not have any sharp edges nor be hot or cold to the touch. For special loads, such as a container of hazardous liquid, or a hospital patient, additional attention should be paid to the lifting process, for instance, by taking special safety precautions and planning the lifting operation.

If a person is to lift loads of a weight unknown in advance to him or her, it is desirable to label the loads in advance showing their weight and possibly advise caution.

Use correct lifting techniques

Sometimes a person can more or less freely choose the lifting technique. In such instances, prior training will ensure that the best possible posture is adopted during the lift. On the other hand, the benefits of information and training should not be overestimated. In practice, improved lifting techniques are often not feasible because of restrictions at the workplace. In addition, ingrained habits and movements can only be changed after intensive training and repetition.

Training should address the following aspects:

- Assess the load and establish where it must be moved to. Consider using the help of others or the use of a lifting accessory.
- Where lifting has to be done without any additional help, stand directly in front of the load. Make sure the feet are in a stable position. Bring the load as close as possible to the body. Grasp the load with both hands, using the whole hand, not just a few fingers.
- Hold the load as close as possible to the body while lifting. Make a flowing movement with a straight trunk. Avoid twisting the trunk. If necessary, move the feet.

The latter recommendation is of great importance in reducing back stress (Figures 2.21 and 2.22). Bending or twisting of the trunk while lifting contributes significantly to injuries of the lower back. The load in Figure 2.21 is approximately 20 kg; bending the trunk forward, as in (a), results in a back stress nearly 30 per cent greater than in (b).

Heavy lifting should be done by several people

Several people can work together if the load is too heavy to be lifted by one person. The partners must be of approximately the same height and strength, and must be able to work well together. One of them must coordinate the lifting, as this will prevent unexpected movements.

Use lifting accessories

Many lifting accessories are available to help lift and move loads. The different types include, for example, levers, raising platforms and cranes. Figure 2.23 gives

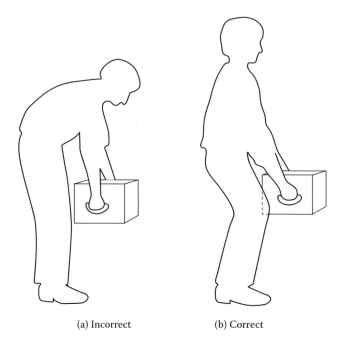

(a) Incorrect   (b) Correct

FIGURE 2.21   Lifting with a bent trunk and a large horizontal distance between load and the lower back (a) is more hazardous than lifting with the back straight and a small horizontal distance between the load and body (b).

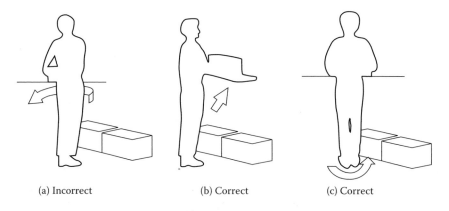

(a) Incorrect   (b) Correct   (c) Correct

FIGURE 2.22   Twisting the trunk while lifting (a) must be avoided by a better choice of lay-down surface (b) or by moving the feet (c).

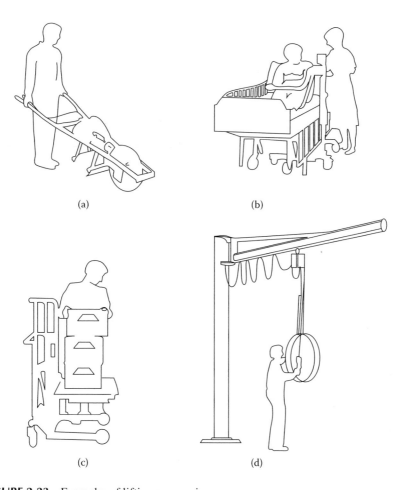

(a)   (b)

(c)   (d)

FIGURE 2.23 Examples of lifting accessories.

an example of a special device for lifting curbstones (a), a dedicated mobile lift for moving patients (b), and two universal accessories: a mobile lifting table (c) and a crane (d).

2.3.2 CARRYING

After a load has been lifted, it must sometimes be moved manually. In general, walking with a load is both mechanically stressful and energetically demanding. As a result of holding the load, the muscles are subjected to continuous mechanical stress; this particularly affects the muscles in the arms and back. Displacing the whole body and the load consumes energy.

Limit the weight of the load

The permissible weight of a carried load is determined mostly by the lifting which precedes the carrying. Guidelines for lifting are given in the section on lifting (p. 29).

Hold the load as close to the body as possible

To limit both mechanical stress and energy consumption, the load must be kept as close as possible to the body. Small, compact loads are therefore preferable to larger loads. By using accessories such as a backpack or a yoke, it is possible to hold the load even closer to the body.

Provide well-designed handgrips

The load should be fitted with well-designed handgrips that have no sharp edges. Alternatively, an accessory such as a hook may be used instead.

Avoid carrying tall loads

A person lifting a tall load will tend to bend the arms to prevent the load from hitting his or her legs. This causes additional fatigue to the muscles in the arms, shoulders and back. The vertical dimension of the load must therefore be limited (Figure 2.24).

Avoid carrying loads with one hand

When only one hand is used to carry a load, the body is subject to an asymmetric stress; well-known examples of this are carrying a school bag, suitcase or shopping bag. The solution is to carry two lighter loads (one in each hand) or use a backpack.

Use transport accessories

There are a large number of provisions and accessories such as roller conveyors, conveyor belts, trolleys and mobile raising platforms which make it unnecessary to

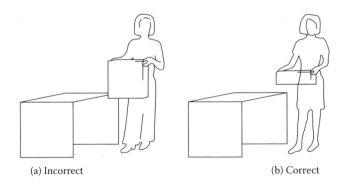

(a) Incorrect   (b) Correct

FIGURE 2.24 Carrying of tall loads should be avoided.

carry loads manually. Whenever one of these is selected, the user must be aware of any possible new problems resulting from the lifting, pulling and pushing required to place the load on the device or move it (trolley, etc.).

2.3.3 PULLING AND PUSHING

Many types of trolleys have to be moved manually. Pulling and pushing trolleys places stress mainly on the arms, shoulders and back. The design of the trolley must take this into account (Figure 2.25).

Limit the pulling and pushing force

When setting a trolley in motion by pulling or pushing, the exerted manual force should not exceed approximately 200 N (about 20 kg-force). Although the maximum possible force required is often considerably higher, this limit should be adhered to in order to prevent large mechanical stresses, mainly to the back. If the trolley is kept moving for more than one minute, the permissible pulling or pushing force drops to 100 N.

In practice this means that trolleys with a total mass (including load) of 700 kg or more should certainly not be displaced manually. The permissible weight depends on the type of trolley, the kind of floor, the wheels, and so forth. Many types of motorized trolleys are available which can be used as an alternative.

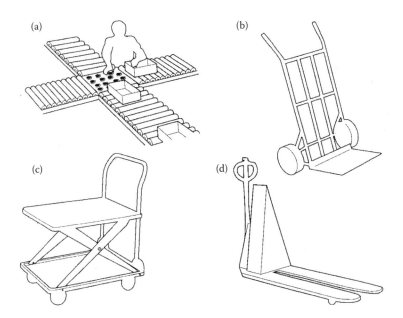

FIGURE 2.25 Transport accessories that replace manual carrying: (a) roller conveyor, (b) sack barrow, (c) mobile raising platform, (d) fork lift.

FIGURE 2.26   Using the weight of the body when pushing or pulling trolleys.

Use the body weight when pulling or pushing

A correct pulling and pushing posture is one which uses the body's own weight. When pushing, the body should be bent forwards and when pulling, it should lean backwards. The friction between the floor and the shoes must be sufficiently large to allow this. There must also be sufficient clearance for the legs to be able to maintain this posture. In pulling and in pushing, the horizontal distance between the rearmost ankle and the hands must be at least 120 cm. When pulling, there must also be room under the trolley to place the forward foot directly below the hands (Figure 2.26).

Provide handgrips on trolleys

Trolleys and such should be fitted with handgrips so that both hands can be fully utilized to exert a force. The dimensions (in cm) of handgrips for pushing and pulling are given in Figure 2.27. The handgrips must be cylindrical. Vertical handgrips, at

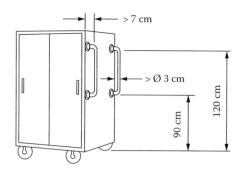

FIGURE 2.27   Recommended handgrip designs for pulling and pushing trolleys.

a height of 90–120 cm, have the advantage that the hands can be placed at the right height when maintaining a correct pulling or pushing posture.

A trolley should have two swivel wheels

Trolleys used on a hard floor surface must be fitted with large, hard wheels, which are able to limit any resistance due to unevenness of the floor. Two swivel wheels should be fitted to achieve good maneuverability. These should be positioned on the side that is pushed or pulled, i.e., where the handgrips are located. Having four swivel wheels is not advisable because this makes it necessary to steer continuously.

The loaded trolley must not be higher than 130 cm so that most persons can see over the load while pulling or pushing.

Ensure that the floors are hard and even

If possible, avoid having to lift trolleys over any raised features such as curbs. If this is unavoidable, the trolleys should be fitted with horizontal handgrips. The weight to be lifted may not exceed the limit given above for lifting (p. 30).

SUMMARY CHECKLIST

BIOMECHANICAL, PHYSIOLOGICAL AND ANTHROPOMETRIC BACKGROUND

Biomechanical Background

1. Are the joints in a neutral position?
2. Is the work held close to the body?
3. Are forward-bending postures avoided?
4. Are twisted trunk postures avoided?
5. Are sudden movements and forces avoided?
6. Is there a variation in postures and movements?
7. Is the duration of any continuous muscular effort limited?
8. Is muscle exhaustion avoided?
9. Are breaks sufficiently short to allow them to be spread over the duration of the task?

Physiological Background

10. Is the energy consumption for each task limited?
11. Is rest taken after heavy work?

Anthropometric Background

12. Has account been taken of differences in body size?
13. Have the right anthropometric tables been used for specific populations?

POSTURE

14. Has a basic posture been selected that fits the job?

Sitting

15. Is sitting alternated with standing and walking?
16. Are the heights of the seat and backrest of the chair adjustable?
17. Is the number of adjustment possibilities limited?
18. Have good seating instructions been provided?
19. Are the specific chair characteristics dependent on the task?
20. Is the work height dependent on the task?
21. Do the heights of the work surface, the seat and the feet correspond?
22. Is a footrest used where the work height is fixed?
23. Are excessive reaches avoided?
24. Is there a sloping work surface for reading tasks?
25. Is there enough legroom?

Standing

26. Is standing alternated with sitting and walking?
27. Is work height dependent on the task?
28. Is the height of the work table adjustable?
29. Has the use of platforms been avoided?
30. Is there enough room for the legs and feet?
31. Are excessive reaches avoided?
32. Is there a sloping work surface for reading tasks?

Change of Posture

33. Has an effort been made to provide a varied task package?
34. Have combined sit–stand workplaces been introduced?
35. Are sitting postures alternated?
36. Is a pedestal stool used once in a while in standing work?

Hand and Arm Postures

37. Has the right model of tool been chosen?
38. Is the tool curved instead of the wrist being bent?
39. Are handheld tools not too heavy?
40. Are tools well maintained?
41. Has attention been paid to the shape of handgrips?
42. Has work above shoulder level been avoided?
43. Has work with the hands behind the body been avoided?

MOVEMENT

LIFTING

44. Have tasks involving manual displacement of loads been limited?
45. Have optimum lifting conditions been achieved?
46. Has care been taken that any one person always lifts less, and preferably much less, than 23 kg?
47. Have lifting situations been assessed using the NIOSH method?
48. Are the weights to be lifted not too light?
49. Are the workplaces suited to lifting activities?
50. Are handgrips fitted to the loads to be lifted?
51. Does the load have a favorable shape?
52. Have good lifting techniques been used?
53. Is more than one person involved in heavy lifting?
54. Are lifting accessories used?

Carrying

55. Is the weight of the load limited?
56. Is the load held as close to the body as possible?
57. Are good handgrips fitted?
58. Is the vertical dimension of the load limited?
59. Is carrying with one hand avoided?
60. Are transport accessories being used?

Pulling and Pushing

61. Are pulling and pushing forces limited?
62. Is the body weight used during pulling and pushing?
63. Are the trolleys fitted with handgrips?
64. Do the trolleys have two swivel wheels?
65. Are the floors hardened and even?

3 Information and Operation

Increasing numbers of people are making use of complex products and systems. This leads to interaction based on receiving information and acting on it. The relationship between information and operation can be represented by a man–machine model (Figure 3.1).

Since computers play a major role in daily life, this chapter will concentrate on working with computers. Most of the principles which are discussed here are not new and certainly do not apply exclusively to working with computers; applications from other fields are given in the examples.

The boundary between man and machine is called the user interface. This chapter will deal with principles for optimization of the user interface by means of better understanding of the user, better presentation of information, better design of controllers for operation and a better interaction between the presented information and the operation (called the dialogue). These principles are applied to website design, mobile interaction, and virtual reality and games.

3.1 THE USER

When designing computer programs (or information systems), it is important to know who the users of the program will be. The possibilities and limitations of these persons determine to a large extent how the user-interface should be designed in an optimal way. The user's limitations are a particularly critical factor because users have to absorb more and more information at higher speeds and complexity. Since ergonomics particularly deals with designing for groups of people, it is important to determine to what group the majority of the users belong and what are the particular characteristics of that group.

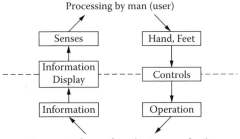

FIGURE 3.1 A man–machine model shows the relationship between information and operation.

Determine the user groups as accurately as possible

When determining the user groups, the following aspects may be important:

- Age
- Nationality (language and culture)
- Ability to read the language (e.g., children)
- Occurrence of special needs within the group (e.g., visually handicapped people)
- Level of education
- Experience with similar tasks and systems
- Frequency with which tasks are performed
- Motivation to perform the tasks
- Physical environment in which the tasks are performed
- Possibilities for education and training
- Existence of other user groups that use only part of the system

Take cultural differences into account

Since users have different cultural backgrounds, it is important to be aware of some intercultural differences. Interfaces for international users have to cope with problems such as these:

- Symbols that are acceptable in one culture may be offensive in another (e.g., symbols of pigs that are considered acceptable in one culture but unclean in others).
- Symbols that are well known in one country may be unrecognizable in another (e.g., symbols for mailboxes that have different shapes and colors in many countries).
- Colors may have different emotional content: red is a happy color in for example China whereas in some other countries it is associated with "danger" or "stop."

For most countries reading from the top left to the bottom right is logical; in other countries reading from the top right to the bottom left is logical.

3.2 INFORMATION

3.2.1 Visual Information

Simultaneous perception of a large amount of information is best achieved by humans through the eyes. This makes the eyes the most important source of information and means that people with only limited eyesight will miss much information or will assimilate it slowly. The form in which the information is presented must be suited to as many people as possible. Perception of information by means of sound and other senses will be discussed shortly.

KEYHOLE   Keyhole

(a) incorrect   (b) correct

FIGURE 3.2 Text consisting entirely of capital letters (a) is not as legible as text consisting of both upper- and lowercase letters (b).

Below are guidelines on the legibility of information presented on screens, in books, in newspapers, etc., and information transferred by overhead sheets, computer presentations, slides, etc.

Do not use text consisting entirely of capitals

In continuous text, lowercase letters are preferable to uppercase letters. The letters with ascenders (b, d, f, h, k, 1, t) and those with descenders (g, j, p, q, y) stand out and contribute to the image of a word (Figure 3.2). The reader can see at a glance what is written and need not read letter by letter. Capitals can be used for the first letter in a sentence, for a title or proper noun and for abbreviations that are familiar to the user.

Use a familiar typeface

Plain characters without much ornamentation are the most legible. In particular headings and captions (directions, names, book titles) a sans serif typeface is preferable to a serif typeface (Figure 3.3).

Avoid confusion between characters

Some characters are difficult to distinguish from each other, which can lead to confusion. The smaller the number of points forming the character, the greater the risk of confusion. In a normal text this will not lead to many difficulties, but in abbreviations and for letters, numbers and code numbers, this could well be confusing. Some frequent causes of confusion are shown in Figure 3.4.

Make sure that the characters are properly sized

The required dimensions of characters depend on the reading distance. A rule of thumb is that the height of capital letters should be at least 1/200th of the reading distance. Letters presented in a conference room 20 meters long should be at least 10 centimeters high on the screen. On computer displays, capitals should be no smaller than 3 mm. The requirements relating to proportions are given in Figure 3.5. It is sensible in the case of height and width of capitals to base these on the capital O and

Serif letters   Sans serif letters

(a) incorrect   (b) correct

FIGURE 3.3 Characters without much ornamentation are most legible.

Mutual	**One-way**
O and Q	C mistaken for G
T and Y	D mistaken for B
S and 5	H mistaken for M or N
I and L	J,T mistaken for I
X and K	K mistaken for R
I and 1	2 mistaken for Z
O and 0 (zero)	B mistaken for R,S or 8

FIGURE 3.4 Avoid confusion between similar characters.

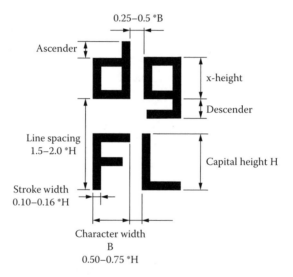

FIGURE 3.5 Use characters of the correct shape.

for lowercase letters, to base these on the lowercase *o*. The stroke width for letters in a text should preferably be based on a nonround letter, for example, the capital *I*.

The longer the line, the greater the required line spacing

In a text, the required distance between lines, that is, the distance between the imaginary lines on which the letters are placed, depends on the length of the line. The line spacing in newspaper columns can therefore be much less than in a book (Figure 3.6). A guideline here is that the line spacing should be at least 1/30th of the line length. If the lines are too closely spaced, it is difficult for the eye to follow from the end of one line to the beginning of the next line.

In a text, the required distance between lines, that is, the distance between the imaginary lines on which the letters are placed, depends on the length of the line. The line spacing in newspaper columns can therefore be much less than in a book (Figure 3.6). A guideline here is that the line spacing should be at least $1/30$th of the line length. If the lines are too closely spaced, the eye finds it difficult to follow on from the end of one line to the beginning of the next.

In a text, the required distance between lines, that is, the distance between the imaginary lines on which the letters are placed, depends on the length of the line. The line spacing in newspaper columns can therefore be much less than in a book (Figure 3.6). A guideline here is that the line spacing should be at least $1/30$th of the line length. If the lines are too closely spaced, the eye finds it difficult to follow on from the end of one line to the beginning of the next.

FIGURE 3.6 Longer lines require a wider line spacing to avoid confusion.

Good contrast contributes to legibility

Whether characters are legible or not depends on contrast, that is, the difference in brightness between the text and the background. Contrast has an even greater influence on legibility than lighting. Some examples are illustrated in Figure 3.7.

For computer displays, other technical characteristics of the screen such as flicker frequency will also play a role. Within a given task, it is not desirable to change frequently from dark symbols on a light background to light symbols on a dark background. For word processing from a draft on paper it is therefore preferable, for example, to use a white screen with black letters. This requires the refreshing rate of the screen to be fairly high to prevent flickering.

Produce diagrams that are easy to understand

Diagrams should be used to support text or as a substitute for text. The use of diagrams has become even more attractive now that computer programs are available which can help those who are not skilled enough to draw them. However, the number

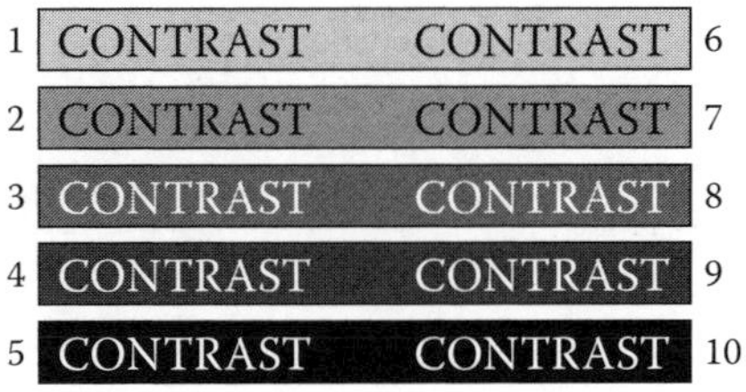

FIGURE 3.7 Choose a good contrast.

of options offered by the computer is, in fact, so large that users with little graphical knowledge may well produce diagrams which are incomprehensible.

Diagrams should be such that everyone can understand them and captions should be legible. Legibility is especially important for computer presentation projectors, and overhead or slide projectors. The correct letter size will enhance legibility (see p. 46). This often requires reducing the amount of text, and therefore the user is encouraged to dispense with all irrelevant information. Here are a few guidelines:

- Titles and captions should not consist entirely of capitals.
- Abbreviations must make sense; do not allow the computer to truncate words at random.
- Different types of shading must be easy to distinguish from one another.
- The scale should match the units being used (e.g., 0–100 for percentages).
- The subdivision of the scale must be sensible.

Use pictograms with care

In principle, symbols are not bound to a particular language. It seems a good idea to replace written signs by symbols, such as pictograms or icons, in public places where many people from different language groups come together. Many of these pictograms are, in actual fact, poorly understood. Here are a few guidelines:

- Bear in mind cultural differences: a picture of a knife and fork can mean a simple snack bar in one country or a luxury restaurant in another.
- People should have a simple mental image of what is meant to be represented by the pictogram. In many people's minds a door can represent either an entrance or an exit; it is therefore difficult to represent these concepts by two different pictograms.
- More stylized pictograms, i.e., further removed from reality, are less well understood.
- Use only pictograms that represent a single concept, not a combination of concepts.

In a popular Internet browser, for example, the icon for reloading the Internet page is not easy to understand without written explanation (Figure 3.8).

3.2.2 HEARING

The ear is not often used in information perception, except in communication through speech. Nevertheless, if the eyes are overtaxed in a particular task, it may

FIGURE 3.8   Icons for an Internet browser.

sometimes be possible to rely on the ears instead. Frequent application of auditory signals is not recommended, however, as even the most pleasant sounds can start to irritate in the long run. The use of speech seems attractive. It is possible to imitate speech by a computer in such a quality that a user does not notice the difference in comparison with a real voice. Sometimes this difference is essential: it determines the way the user reacts. An example is computer speech via the telephone (voice-response systems).

Sound should be reserved for warning signals

The ear is particularly suited to detecting warning signals because it is difficult for people to escape from sounds that come from all directions. In contrast to a warning given by a light signal, a person does not have to be at a particular spot to detect a warning sound. If a task consists of noticing when limit values are exceeded, the number of missed signals will increase with time if vision is the only sense used, but the task will be easier if a bell or buzzer sounds when the limit is exceeded.

Select the correct pitch

If the distance between the device and the operator is large, the sound intensity must be large and the pitch not too high. Low pitches are more suitable if the sound has to travel around corners and obstacles.

The signal must be clearly distinguishable from background noises, as is the case, for example, with sirens on fire engines and police cars. A high-pitched signal is better if the background includes many low-pitched sounds. The signal must be regularly interrupted where there is continuous background noise.

Synthesized speech must have adjustable features

It is easier for a computer to synthesize speech than to recognize speech. The user should be able to control computer speech in the following ways:

- Alter its speed (faster or slower)
- Produce a repeat of any section
- Interrupt (pause)
- Switch over to another form of display

Attempts to make computer speech sound human are not always desirable. In many instances the user wants to be able to hear the difference between a computer and a person.

3.2.3 OTHER SENSES

Humans possess three other senses apart from sight and hearing: smell, taste and touch. These senses can also be used as sources of information.

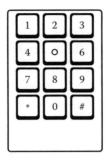

FIGURE 3.9   Make positions on a keyboard perceptible to touch.

Restrict the use of taste, smell and temperature to warning signals

Taste, smell and temperature should be used only to indicate alarm conditions; an example is the addition of odor to natural gas. These senses should be used only sparingly. Furthermore, they are inadequate for multiple use. For example, two smells present together cannot easily be distinguished from each other.

Use the sense of touch for feedback from controls

The sense of touch can be used to provide feedback on the location and status of controls. It is particularly useful in places outside the visual field. Examples of this are the bass buttons on an accordion or the identifying marks on the 5 of a numerical keypad (Figure 3.9). Haptic interfaces, combining touch and force feedback, are becoming popular in games and also in industrial appliances, such as in operating-rooms.

Use different senses for simultaneous information

Modern information technology allows the simultaneous presentation of a large amount of information. If the information comes through different senses, this multiplicity will present no problem: driving a car while the radio is on is no more difficult than with the radio off, but reading a book when observing a control panel is more difficult because the same sense organs are being used together. On the other hand, if variations or alarm states are conveyed by a warning bleep, reading can indeed take place while controlling. It is recommended that important information such as alarms be conveyed via several senses simultaneously, named multimodal interaction. A good example is a sound coupled with a light signal. An example where many senses are simultaneously used is the use of cell phones while driving cars: the amount of information to digest and the load of the memory are both very high.

3.3   CONTROLS FOR OPERATION

People transmit their decisions to machines by means of controls, such as knobs, handles or steering wheels. Computer systems are controlled mostly via keyboards, mice, touch pads and joysticks. In games and computer simulations nowadays hands-free controls are becoming popular. In this chapter we discuss how controls operate.

There are two types of controls, ones with a fixed position that are generally directly fixed to the information device (fixed controls), and ones that can be operated from a distance (wireless, remote, and hands-free controls).

3.3.1 FIXED CONTROLS

Controls must be given easily distinguishable shapes whenever the use of the wrong control could have undesirable consequences.

Use familiar keyboard layouts

For keyboards the QWERTY layout has been in use for a long time. In this layout, the top row of letters starts with the letters Q-W-E-R-T-Y. The most important advantage of this layout is its worldwide distribution and acceptance as a standard. For numerical keypads there are three international standards for the layout of a numerical keypad: the standard for pocket calculators with 1-2-3 on the bottom row, the standard for push-button telephones with 1-2-3 on the top row, and a layout where all keys are in a line, such as in the top row of keys on a keyboard (Figure 3.10). The choice of the most suitable layout for a particular task should take into consideration which of the three layouts is the most logical or least confusing.

Restrict the number of function keys

Function keys can be used to avoid typing system commands letter by letter. However, the meaning of a function key must be clear under all circumstances. It is advisable to distinguish function keys from other keys through size, shape, color or position. Function keys of similar shape are best grouped in sets of three or four, because it is quite easy to select the outer key in a row or the one next to it. There are three types of function keys:

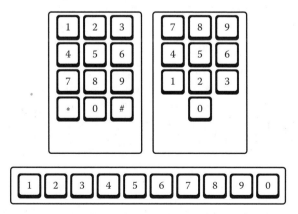

FIGURE 3.10   Select a standard layout for numerical keypads, one that does not cause confusion.

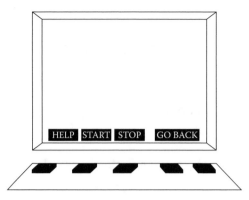

FIGURE 3.11   When using softkeys, the meaning of the function keys must be clear.

- Hard (independent) function keys, which have a fixed meaning, irrespective of application. Familiar examples include the cursor keys. They can also be used for the following purposes: start/stop (on/off), help, do (execute), restore, delete, and print.
- Programmable function keys, where the user determines the meaning. This could be, say, a frequently used sequence of system commands.
- Variable function keys, where the meaning depends on the current application and is determined by the software. In this case, it is sensible for the meaning to appear at a corresponding position (Figure 3.11).

Match the type of cursor control to the task

The best method of pointing out something on a screen in a given situation depends, among other things, on the required speed, accuracy and direction of movement. Input can be with a joystick, the user's fingers, a light pen, function keys or a mouse. When dealing with text, cursor keys are adequate for indicating position, but their relative location on the keyboard must correspond to the direction of movement (Figure 3.12).

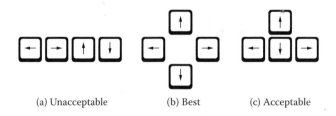

FIGURE 3.12   The relative position of cursor keys must correspond to the direction of movement.

Do not use the mouse exclusively

Frequent use of the mouse can cause an overload and one-sided load of certain muscles and tendons in hands and wrists (RSI). It is therefore recommended to alternate the use of the mouse with the use of the keyboard, for instance, by using so-called short-cuts. These are combinations of keys, e.g., the Ctrl key and the C to copy, or typing the first (underlined) letter of a menu item to select it.

Touch screens are suitable for inexperienced users

Touch screens offer a suitable form of dialogue with a computer if inexperienced, casual users need to retrieve information from databases (Figure 3.13). An advantage of this method is the direct relationship between what the eyes see and what the hands do. The disadvantages are that:

- The arms are often in an uncomfortable, stretched position.
- The areas to be selected on the screen need to be relatively large.
- The user must be close to the screen.
- The screen becomes dirty.

Use pedals only if the use of the hands is inconvenient

Pedals are an option whenever a large force is to be exerted or if both hands are otherwise needed for precise control. The number of pedals in any situation should not exceed three. The position of both feet should not differ too much while operating the pedals. Frequent use of pedals should be avoided in activities that are carried out while standing.

Avoid unintentional operation

Touching a control unintentionally, for example with the hand or arm, could change its state without anyone noticing. In some instances, the consequences could be disastrous. Unintentional operation can be avoided by adjusting the mode of opera-

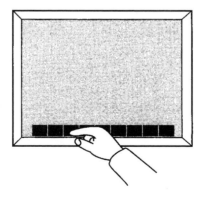

FIGURE 3.13   Touch screens will help inexperienced users in searching.

tion (rotating knob instead of push button) or the location of the control (countersunk into the keyboard). The risk of unintentional operation exists with:

- Handles that can be moved in different ways, each having a different effect (e.g., a combination of turning indicator and light switch in a car).
- Keys, which do not have to be pressed in but only touched, so-called fingertip controls.

The risk of confusion is much greater if the controls are closely spaced. In many cases these should be replaced by so-called membrane keys, which have to be pressed in to some degree. The use of fingertip controls and membrane keys must be limited to locations where dirt could impair the proper operation of mechanical keys (controls in digital metal-working machines) or where there are strict hygiene requirements (food industry, medical equipment).

Think carefully before using labels and symbols

The use of labels or symbols on controls may seem a good idea. The number of possibilities is considerable, but the prerequisites for use are:

- Sufficient space
- Sufficient lighting
- Adequate letter size
- Location close to or on the control
- Use of straightforward, everyday words
- Use of readily understood symbols
- Information being restricted to the meaning of the control

A familiar example of the use of labels is on a keyboard. Although mostly capital letters appear on the keys, lowercase letters appear on the screen.

Limit the use of color

Although the eye can distinguish between a large number of colors, it is advisable to use only the following five colors for color discrimination of controls: red, orange, yellow, green and blue. Five points should be kept in mind:

- The difference with respect to the background color, and the contrast
- The association that Western people make with some colors (red for danger, green for safe)
- Reduced color discrimination (color blindness)
- The color and lighting of the surroundings
- Color strongly attracts attention; its use should therefore be limited

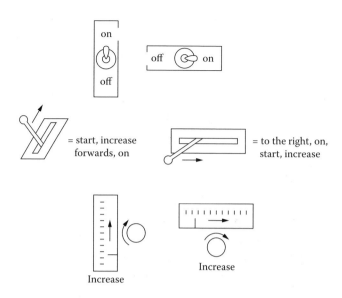

FIGURE 3.14 Movement of controls must correspond to display movements.

Ensure compatibility in the direction of movement

The movements on a display must correspond to those of the controls (Figure 3.14). The following actions are the most natural in achieving a desired result:

On: upward, to the right, away from the user, clockwise, pulled out
Off: downward, to the left, toward the user, anticlockwise, pressed in
Increase: upward, to the right, away from the user, clockwise, increasing resistance
Decrease: downward, to the left, toward the user, anticlockwise, decreasing resistance

There are exceptions where the relationship is ambiguous. These are mostly due to ingrained habits. The direction of motion on a brake pedal in a car, for example, represents increasing brake force, but also decreasing speed.

The objective of a control must be obvious from its location

The location of a control must bear a logical relationship to the position of the relevant information or to the effect. Softkeys, whose function is displayed on the screen directly above the keys, have already been discussed.

Another example is a stove, where the relationship between the location of the control knobs and that of the burners can be made to correspond, by placing the knobs on the top of the stove or next to the burners in the same layout.

3.3.2 Wireless, Remote and Hands-Free Controls

The traditional keyboard and mouse are now also available wireless. That means that the user can move freely around the information systems or displays. The functionality is similar to that of the traditional keyboard and mouse.

Remote controls can also offer new functionalities. Most remote controls have a large number of small buttons with little icons or pictograms representing the function. They are commonly used to operate electrical equipment, and the number of systems with remote controls is growing. A new generation of remote controls can be operated by means of gesture, gaze or speech, without contact between the body and the system ("hands free").

**Use wireless, remote and hands-free controls
when the user needs more freedom**

The use of remote controls has advantages if the user cannot be, or does not wish to be, in direct contact with the system being controlled. Examples are industrial cranes and home video or DVD recorders. Guidelines for the use of wireless, remote and hands-free controls include the following:

- The design must assure that the direction of operation is evident (see Figure 3.15).
- If several systems have to be operated remotely from a single point, only one remote control must be used.
- No more than two systems should be controlled simultaneously.
- Important function keys (e.g., on/off) must have a fixed location, for example, in the upper right-hand corner.

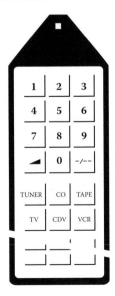

FIGURE 3.15 The shape of the remote control must indicate clearly how it should be held.

Do not use hands-free controls for precise data

Hands-free controls by gesture, gaze or speech are useful in situations where people cannot use their hands because the hands are needed for other tasks or because of disability. The main problems with hands-free controls are the lack of sufficient error correction and the acceptance by the computer of only a limit set of input signals.

Use speech recognition in quiet environments

In speech recognition the user talks to the computer and the computer recognizes the speech. Speech recognition is only acceptable in quiet environments or when the user uses a microphone. The extent to which the computer recognizes the vocabulary is still limited.

3.4   DIALOGUES

A user and a system interact: after an operation (action) of the user, the system reacts, which results in a reaction of the user and so on. For effective interaction between user and system the user friendliness of the system and the form of the two-way communication (dialogue) are important.

3.4.1   USER FRIENDLINESS

Several principles are formulated to promote a user-centered approach to the development and evaluation of a product or system, so the use of computerized interactive tools is enhanced by the improved quality of the dialogue. The aim is to increase the ease of use of the dialogue in terms of effectiveness, efficiency and satisfaction.

Make the dialogue suitable for the user's task and skill level

A dialogue is suitable for the task to the extent that it supports the user in the effective and efficient completion of the task. Some typical application requirements of this principle are:

- The system should present the user with only those concepts which are related to the user's activities in the context of the task.
- Any activities required by the system but not related to the user's task should be carried out by the system itself.
- The type and format of input and output should be specified such that they suit the given task.

Make clear in the dialogue what the user should do next

In a clear dialogue each dialogue step is immediately comprehensible through feedback from the system or is explained to the user when he or she requests the relevant information. Some typical applications of this principle are:

Graphics: Horizontal Line

1 – Horizontal position	Full
2 – Vertical position	Baseline
3 – Length of line	
4 – Width of line	0.013
5 – Grey shading (5% of black)	100%

Selection: 0

FIGURE 3.16 The system must provide hints to the user.

- After any user action, the system should have the capability to initiate feedback.
- Feedback or explanations should assist the user in gaining general understanding of the dialogue system.
- If defaults exist for a given task, they should be made available to the user, for instance, by giving hints (Figure 3.16).

Give the user control over the pace and sequence of the interaction

A dialogue is controllable when the user can control the whole course of the interaction until the point at which the goal has been met. Some typical applications of this principle are:

- The speed of the operation should not be dictated by the system.
- If interactions are reversible and the task permits, it should be possible to undo the last dialogue step.
- The way that input/output data are represented should be under the control of the user, thus avoiding unnecessary input/output activities such as entering 00123 when just 123 is more reasonable.

Make the dialogue consistent

A dialogue is consistent when it corresponds to the user's task knowledge, education, and experience and to commonly accepted conventions. Some typical applications of this principle are:

- Dialogue behavior and appearance within a system should be consistent; the end of a command, for instance, should always occur in the same manner, for example, with "enter" or "return" or simply nothing.
- The application should use terms, which are familiar to the user in the performance of the task; languages, for example, English and Dutch should not be mixed (Figure 3.17).

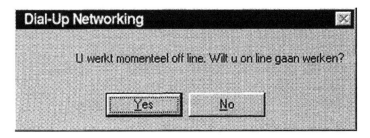

FIGURE 3.17   Using English programs within a Dutch operation system leads to bilingual dialogues.

Make the dialogue forgiving

A dialogue is forgiving when, despite evident errors in input, the intended result may be achieved with either no or minimal corrective action having to be taken. Typical applications of this principle are:

- Errors should be explained to help the user to correct them.
- The application should prevent the user from making errors.
- Error messages should be formulated and presented in a comprehensible, objective and constructive style and in a consistent structure. Error messages should not contain any value judgments, such as "this input is nonsense."

Make the dialogue customized to suit the user

A dialogue can be customized to suit the user when the system is constructed to allow for modification to the user's individual needs and skills for a given task. Some typical applications of this principle are:

- The amount of explanation (e.g., details in error messages, help information) should be modifiable according to the individual level of knowledge of the user.
- The user should be allowed to incorporate his or her own vocabulary to establish individual naming for objects and actions.
- The user should be able to modify the dialogue speed to his or her own working speed (for instance, the speed of scrolling information).

Make the dialogue suitable for learning

A dialogue is suitable for learning to the extent that it provides means, guidance and stimulation to the user during the learning phases. Some typical applications of this principle are:

Did you know this?

You can select more files or maps
by keeping the CTRL pressed, while
you click on the next files or maps

FIGURE 3.18   The "tip of the day" helps the user understand the program.

- Help information should be task dependent.
- All kinds of strategies which help the user to become familiar with the
  dialogue elements should be applied, for instance, standard locations of
  messages and similar layout of screen elements. An example is the tip of
  the day in some programs (Figure 3.18).

3.4.2   Different Forms of Dialogue

The way in which the exchange of information between people and computers is
implemented in the software constitutes the dialogue form.

Not only computers are equipped with a display. Increasing numbers of other
types of industrial systems include a small screen, sometimes with only a few lines
(e.g., a mobile phone). Therefore, the rules described below for dialogue forms, with
their advantages and disadvantages, are also applicable to such systems. The suitabil-
ity of a given form depends on the application. A dialogue that is suitable for learning, such as described above, usually includes a combination of dialogue forms.

Use menus for users with limited knowledge or experience

With a menu, the user selects from a list of alternatives (Figure 3.19). The great
advantage is that the user needs little knowledge or experience to understand the
system. Furthermore, this type of dialogue requires little typing, it requires little

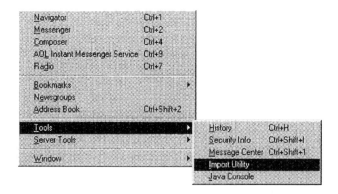

FIGURE 3.19   Menus do not require great mental effort on the part of the user.

mental effort, and if a task is interrupted, the system immediately gives instructions on how to continue. The major disadvantage is that experienced users find it fairly long-winded and slow. In addition, the user has to read too much irrelevant information. Selections are made more rapidly from menus that have approximately seven topics.

If the menu does not fit on a page, a hierarchy, comparable to a family tree, has to be created. Because this means that only a small portion of the hierarchy is ever shown on the screen, there is a chance that the user will lose his or her way. Complex hierarchies must be avoided.

Recognize the limitations of an input form

An input form consists of a number of filled-in screen sections (protected fields) that highlight the questions, together with some sections (free fields) to be filled in by the user to answer the questions. Here too, little training is required, as many users are familiar with paper form filling. Just as with a menu structure, it requires little mental effort on the part of the user. The difficult manipulation of the cursor is a disadvantage because it may not be placed on the protected fields. What is more, the dialogue is not flexible: data must usually be entered in a fixed sequence.

Restrict the use of command language to experienced users

In command language, the user must key in a fixed combination of characters to make the system react. The number of actions by the user is relatively small. To achieve the objective, the user therefore uses cryptic terms, which are meaningless to a layman, for example CA.

The user must know the effect of the various input commands and which commands are required to achieve a given effect, both of which require considerable mental effort.

Direct manipulation must be consistent

A modern, frequently used form of dialogue is direct manipulation. This form is characterized by visual objects on the screen, where the user can point at them, remove them or change their value. A large part of the procedure with Apple Macintosh or Microsoft Windows is based on this form of dialogue. Sometimes the method is inconsistent, such as in the start menu of Windows, where verbs and nouns are mixed.

Recognize the advantages and disadvantages of natural language

When controlling by natural language, the user communicates with the computer in the same manner as in normal conversation with people. This is the most familiar form of communication and gives the user a large degree of freedom. Unfortunately, the disadvantages can often outweigh the advantages. Good understanding between people and computers requires that they speak the same language (for example, Standard English). Natural language not only requires many more words than common language (Figure 3.20), but also natural language is often vague and ambiguous. For

```
Computer: What train do you want?
User: I want to go to London this evening.
Computer: Can you repeat the destination?
User: London.
Computer: From Liverpool to London?
User: No.
Computer: Where do you want to depart from?
User: Bristol.
Computer: From Bristol to London?
User: Yes.
```

FIGURE 3.20 Speech dialogues require additional feedback.

example, in the sentence "The man hit the girl with the umbrella," it is not obvious who was carrying the umbrella. Natural language can make the user overconfident in the system: he understands this, and then he must be able to do more.

3.4.3 HELP

The user's need for help is usually considerable because he or she did not design the (often complex) products and systems. Many manufacturers try to obviate this by supplying bulky instruction manuals along with their equipment. However, this does not appear to be the correct solution, as the manuals are seldom easily accessible and are often mislaid. Help should therefore be provided not only in paper form but also, wherever possible, on the system itself.

Match the type of help to the experience of the user

To be effective, the type of help must be suited to the experience of the user. Differences between the help requirements of experienced users, such as programmers, and inexperienced users are shown in Table 3.1. Different types of help must be offered in parallel because systems are seldom accessed by only one type of user.

TABLE 3.1
Help Requirements Depend on the User

Experienced user	Inexperienced user
Help commands	Help commands and keys
Computer jargon	No jargon, or else additional explanation
Mathematical notation	Examples
Error messages	Suggestions for corrections
No positive feedback	Positive feedback
Screen full of help	Screen position also visible in addition to help
Computer-oriented help	User-oriented help

3.5 WEBSITE DESIGN

More and more information is available through websites on the Internet or Intranet (internal network in an organization). The tools for designing websites are widely available, and therefore, many organizations and individuals make websites. Below some general guidelines for the design of websites are presented.

Due to the rapidly emerging web technology, it is recommended to use technology-specific guidelines for website design presented in recently published books, or published on the Internet by the World Wide Web Consortium (W3C).

Predict the behavior of the user on the site

One best way to explore how a site will be used is to ask novice users (non experts) to try to obtain information or to perform a certain action such as ordering a product. Then the following events may happen:

- Use of the website in an unintended way
- Difficulties in reading or comprehending text
- Difficulties in using regular controls such as keyboard or mouse properly
- Availability of only small displays or a slow Internet connection
- Difficulties in the language in which the web pages are written
- Improper environment for the proper use of eyes, ears or hands
- An old version of the browser

In order to reach the intended user group the content must be easy to access for that user group. Two main principles that contribute to web site accessibility are described below as guidelines.

Make websites accessible by means of different types of hardware and software

Not all users have the facilities and capacities to use and interpret every website format. The content of the website should always remain intact regardless of the technical facilities of the user. Therefore a website should not be designed in such a way that it can be used with only a particular program, technical facility or specific browser. Guidelines to facilitate general accessible websites are:

- Auditory and visual content must be easily convertible into text, also to facilitate blind and deaf users.
- Coding information by only color must be avoided (to facilitate colorblind users and to avoid colors being represented in a different way on other hardware).
- Proper natural language should be used, no strange abbreviations.
- Tables should be possible to convert easily to other formats.
- When using new technologies (e.g., JAVA applets), they should be able to be converted to other systems or be removed, i.e., they should not contain the most important content.

- Windows should be scalable.
- Short sentences should be used and text should be divided into paragraphs to enhance reading from small displays.

Make content understandable and navigable

Not only should the language be clear and simple, but the website should also provide understandable mechanisms for navigating (or browsing) within and between pages. It is hard for users to maintain the overview when they can view only a portion of a page because they are accessing the page either one word at a time (speech synthesis or Braille display) or one section at a time (small display or a magnified display). Some important factors to make content understandable are:

- Consistent navigation methods (e.g., the back button of the browser always goes back to the last visited page)
- Proper names of pages and site maps to show the total structure of the site
- Appropriate use of mechanisms for coding information (bold, italic, color, underlining, words in capitals, etc.)

3.6   MOBILE INTERACTION

Communication between people and between persons and machines is more and more done by means of mobile interaction. The main applications include mobile telephones, handheld computers and car navigation systems.

Use mobile interaction for simple content

The common property of mobile interaction devices is the combination of a small screen, a tiny keyboard and other controls such as voice and pens. Users have little trouble when reading or interacting with simple content, that is, short chunks of material with limited navigation. For more complex functions or larger content interaction design needs more attention.

Adapt screen layout and navigation to the needs of the users

When designing the interaction for mobile access, the following guidelines must be applied:

- Make the interaction simple; not all users want to browse through large amounts of information.
- Design the content for optimal use in the mobile device.
- Make clear when the location of use is essential (e.g., for wireless systems or global positioning systems).
- Make a clear distinction between locally stored data and data on a remote server.
- Make sure that essential information is achieved within a few steps.

3.7 VIRTUAL REALITY

Virtual reality is an environment that is generated by a computer. For example, virtual reality can be used for designing a building or a public space. When virtual reality is combined with the real world, the entire system is called augmented reality; for example, in flight simulators real pilots are sitting in a projected virtual environment. Computer games are other applications of virtual reality. Guidelines for virtual reality relate to the way of manipulating objects in the environment and the usability. The usability of games is also called playability.

Choose an appropriate way of manipulating objects

There are different methods for manipulating objects in virtual environments, each with its own characteristics. The choice depends on the intended similarity with the real world.

- Direct user control works with gestures that mimic real world interactions, for instance, when playing virtual volleyball, the movements of the hands and the body should represent the real world game.
- Physical control uses devices the user can physically touch, such as the steering handle in a flight simulator.
- Virtual control offers devices that the user can virtually touch, this is useful in augmented worlds, for instance, to remove virtual furniture in interior design.
- Agent control works with commands to an entity in the virtual world. Agent control is useful to perform complex operations such as calculations or quick searches or to perform actions that are not possible in the real world such as multiplying objects.

Ensure the playability of a game by enlarging the user experience

The goal of the game is to have fun. Usability is not enough to reach fun. The virtual reality must also provide emotional experiences. Some guidelines for creating emotional experiences are:

- Create actions based on expectations of the user (both what the user expects and what the user does not expects).
- Make the first impression not too complex because it is used to give meaning to observed actions.
- Relate things that happen to desires, hopes and fears.
- Give opportunities for reflection: evaluating and making judgments on the progress of the game.
- Relate events to the user's personal history within the game and the user's hope for the future.
- Relate experiences to each other.

SUMMARY CHECKLIST

THE USER

1. Is the user population defined as accurately as possible?
2. Are cultural differences taken into account?

INFORMATION

Visual Information

3. Have texts with only capitals been avoided?
4. Have familiar typefaces been chosen?
5. Has confusion between characters been avoided?
6. Has the correct character size been chosen?
7. Are longer lines more widely spaced?
8. Is the contrast good?
9. Are the diagrams easily understood?
10. Have pictograms been properly used?

Hearing

11. Are the uses of sounds reserved for warning signals?
12. Has the correct pitch been chosen?
13. Is synthesized speech adjustable?

Other Senses

14. Are the uses of taste, smell and temperature restricted to warning signals?
15. Is the sense of touch used for feedback from controls?
16. Are different senses used for simultaneous information?

CONTROLS FOR OPERATION

Fixed Controls

17. Is a familiar keyboard layout used?
18. Is the number of function keys limited?
19. Is the type of cursor control suited to the task?
20. Is the mouse not used exclusively?
21. Are touch screens used to facilitate operation by inexperienced users?
22. Are pedals used only where the use of the hands is inconvenient?
23. Is unintentional operation avoided?
24. Are labels or symbols properly used?
25. Is the use of color limited?
26. Is the direction of movement consistent with expectation?
27. Is the objective clear from the position of the controls?

Wireless, Remote and Hands-Free Controls

28. Are remote controls used to give the user more freedom?
29. Are hands-free controls avoided for precise data?
30. Is speech recognition limited to quiet environments?

DIALOGUES

User Friendliness

31. Is the dialogue suitable to the user's task and skill level?
32. Does the dialogue make clear what the user should do next?
33. Will the user be in control of the pace and sequence dialogue?
34. Is the dialogue consistent?
35. Is the dialogue forgiving?
36. Is the dialogue suitable for individualization?
37. Is the dialogue suitable for learning?

Different Forms of Dialogue

38. Have menus been used for users with little knowledge or experience?
39. Are the limitations of an input form known?
40. Has command language been restricted to experienced users?
41. Is direct manipulation consistent?
42. Have the disadvantages of natural language been recognized?

Help

43. Is the type of help suited to the level of the user?

WEBSITE DESIGN

44. Can the behavior of the user on the website be predicted?
45. Is the presentation of the website independent of the user's hardware and browser?
46. Is the content understandable and navigable?

MOBILE INTERACTION

47. Is mobile interaction used for simple content?
48. Are screen layout and navigation adapted to the needs of the user?

VIRTUAL REALITY

49. Is an appropriate way of manipulating objects chosen?
50. Is the playability of a game improved by enlarging the user experience?

4 Environmental Factors

Physical and chemical environmental factors such as noise, vibration, lighting, climate and chemical substances can affect people's safety, health, comfort and performance.

In this chapter we deal with these five factors in turn. Other environmental factors such as radiation and microbiological pollution (e.g., bacteria, molds) are not discussed in this book. Guidelines on the maximum allowable exposure are given for each of these factors, followed by possible measures for reducing the exposure. In general three types of measure can be applied to reduce or eliminate the adverse effects of environmental factors:

- At source (eliminate or reduce source)
- In the transmission between source and man (isolate source and/or man)
- At the individual level (reduction of exposure duration, personal protective equipment)

4.1 NOISE

The presence of high noise levels during a task can be annoying and, in time, result in impaired hearing. The first symptom of impaired hearing is a perceived difficulty in understanding speech in a noisy environment (party, pub, etc.). Annoyance, such as interference in communication or reduction of concentration, can occur even at relatively low noise levels. Annoyance and impaired hearing can be avoided by setting upper limits for noise levels. Noise levels are expressed in decibels, dB(A). Table 4.1 illustrates a few examples of noise levels.

4.1.1 GUIDELINES ON NOISE

The guidelines on noise given in this section relate to the prevention of damage to hearing, as well as to the limitation of annoyance.

Keep the noise level below 80 decibels

A noise level that, over an 8-hour working day, exceeds 80 dB(A) on average can damage hearing. Assuming constant noise levels, this daily level will be reached, for example, with an 8-hour exposure to 80 dB(A), or with a 1-hour exposure to 89 dB(A) (Figure 4.1).

Doubling the exposure time requires lowering the permissible noise level by 3 dB(A). If noise levels are variable, the average daily level is calculated from the indi-

TABLE 4.1
Examples of Noise Levels

Type of Noise	dB(A)
Jet engine at 25 m	130
Jet aircraft starting at 50 m distance	120
Pop group	110
Pneumatic chisel	100
Shouting, understandable only at short distances	90
Normal conversation, understandable at 0.5 m	80
Radio playing at full volume; loud conversation	70
Group conversation	60
Radio playing quietly; quiet conversation between two people	50
Library reading room	40
Slight domestic noises	30
Gentle rustling of leaves; indoors in the evening in a quiet neighborhood	20
Very quiet environment	10
Threshold of hearing	0

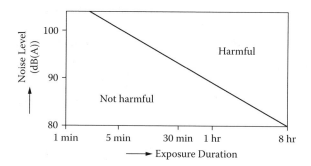

FIGURE 4.1 Constant noise levels should not last too long. To prevent impaired hearing, the average daily level should not exceed 80 dB(A).

vidual noise levels. It is recommended that machines and such be designed to keep the noise level at any time below 80 dB(A).

Avoid annoyance

Annoyance during thinking and communication tasks can arise at levels well below 80 dB(A). An excess of noise will prove annoying even though the limit for damage to hearing has not yet been reached. It is mainly noise produced by others, unexpected noise and high frequency noise that cause annoyance. Table 4.2 provides guidelines for the maximum allowable noise levels to avoid annoyance during various activities.

TABLE 4.2
Maximum Noise Levels to Avoid Annoyance during Various Activities

Activity	dB(A)
Unskilled physical work (e.g., cleaning)	80
Skilled physical work (e.g., garage work)	75
Precision physical work (e.g., fitting and turning)	70
Routine administrative work (not full-time occupation)	70
Physical work with high precision requirements (e.g., fine grinding)	60
Simple administrative work with communication (e.g., activities in typing pools)	60
Administrative work with intellectual content (drawing and design work)	55
Concentrated intellectual work (e.g., working in office)	45
Concentrated intellectual work (e.g., reading in library)	35

Rooms should not be too quiet

Although the aim is to reduce noise levels to below a certain maximum, at the same time, the level should not drop below 30 dB(A), otherwise unexpected irrelevant noise becomes too obvious.

4.1.2 NOISE REDUCTION AT SOURCE

The most fundamental measures in noise reduction are ones taken at source. A number of possibilities for achieving this are discussed below.

Choose a low-noise working method

Consideration should also be given to noise levels when selecting a particular working method. A less noisy working method is not only of importance to those exposed to the noise; in many cases it also means less machine wear and less damage to the product. It is sometimes also possible to reduce or eliminate certain noisy stages in a process, for example, finishing off by grinding can sometimes be partly or totally eliminated.

Use quiet machines

Developments in the construction of "quiet" equipment have resulted in the availability of an increasing number of quiet machines, tools and accessories. When selecting machines for purchase, attention should be paid to potential noise production during normal usage.

Maintain machines regularly

Poor fit, eccentricity and imbalance cause vibration, wear and noise. Regular maintenance of machines and equipment is therefore of great importance.

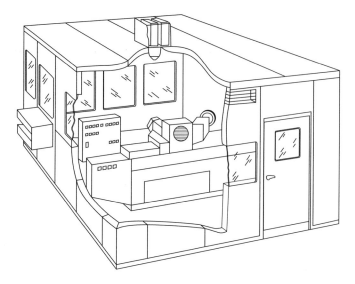

FIGURE 4.2 Noisy machines such as woodworking machines can be enclosed to reduce noise in the vicinity.

Enclose noisy machines

Noisy machines can be placed in a sound-insulating enclosure (Figure 4.2). This can significantly reduce noise levels, but the disadvantage is that enclosed machines are less accessible for operation and maintenance. Special arrangements are also required to bring in and remove any process material, and possibly also to provide ventilation.

4.1.3   NOISE REDUCTION THROUGH WORKPLACE DESIGN AND WORK ORGANIZATION

Noise reduction is achieved in most cases by reducing or preventing the transmission of noise between source and receiver. A few relevant measures to improve the layout of the workplace and work organization are given in this section.

Separate noisy work from quiet work

Noisy activities can be segregated from quiet activities by having them carried out in separate areas or outside normal working hours. The advantage is that fewer people are exposed to the noise, but other measures still have to be taken to protect those who are nevertheless exposed to the noise.

Keep an adequate distance from the source of noise

The correct choice in selecting a location for a noise source is to keep it as far away as possible from those who may be exposed to it. Increasing the distance is most effective close to the source. For example, a given increase in distance, say, of 5 m, from 5 to 10 m has more effect than an equal increase from 20 to 25 m.

Use the ceiling to absorb noise

The ceiling is often used to absorb noise. Although this decreases the noise level only to a limited extent, it is particularly effective in reducing annoying effects such as echo. Measures involving the ceiling are worth undertaking mainly in rooms where sound reverberates and where many workers are present. In existing buildings or in localized applications, one can use loose elements made of sound damping material, which are hung from the ceiling. Another possibility is to install a lower ceiling made of such material. This also offers a way of concealing pipe work, ducting, leads and such, and can help thermal insulation.

Use acoustic screens

Acoustic screens placed between the source and the person can reduce the noise level. This measure is often meaningful only in combination with a sound-absorbing ceiling. The screen should be large enough to prevent the source of noise from being seen. Acoustic screens are ineffective if the distance between the source and the person is large.

Various types of screens are available: as a fixed wall, a moveable screen, a screen hung from the ceiling or one that can be attached to a machine.

4.1.4 Hearing Conservation

One can resort to protecting hearing by using earplugs or earmuffs if the previous measures, which were aimed at the source or at the transmission, are not feasible. Ear protectors must be available if the noise level is temporarily too high, for example during noisy maintenance activities. Different types of ear protectors are illustrated in Figure 4.3.

Earplugs are fitted into the ear, which means that the degree of noise reduction is often limited if they are not properly used. Earmuffs, by contrast, are placed over the ears. The resulting noise reduction is often greater than with plugs. They are also

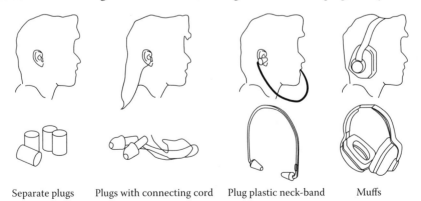

Separate plugs   Plugs with connecting cord   Plug plastic neck-band   Muffs

FIGURE 4.3 Examples of ear protectors. Ear protectors should be used only if other methods have failed to reduce noise sufficiently.

more convenient for frequent donning and removing, and they are more hygienic. Many people find them uncomfortable to wear because of perspiration, and they are also less suitable for those who wear spectacles. Noise reduction will be limited if the muffs do not fit closely over the ears.

Hearing conservation measures must be suited to the noise and to the user

The pitch (frequency) of the noise must be taken into account when choosing ear protectors. Different types of protective equipment have maximum damping effects in certain frequency ranges. Data on the characteristics of ear protectors can be obtained from the suppliers. In order to encourage the use of ear protectors, personal preferences in comfort and ease of use must be taken into account. Different types of ear protectors should therefore be available.

4.2   VIBRATION

In any discussion of vibration, a distinction has to be made between whole-body vibration and hand–arm vibration. In whole-body vibration, the whole body is brought into vibration via the feet (in standing work) or via the seat (in seated work). Usually, the vibration is predominantly vertical, such as in vehicles. Hand–arm vibration affects only the hands and arms, and often arises when using motorized handheld tools.

Three variables are important in assessing vibrations: their level (expressed in m/s), their frequency (expressed in Hz) and the exposure duration. Low-frequency body vibrations (<1 Hz) can produce a feeling of seasickness. Body vibrations between 1 and 100Hz, especially between 4 and 8 Hz, can lead to chest pains, difficulties in breathing, low back pain and impaired vision.

The possible consequences of hand–arm vibration frequencies between 8 and 1000 Hz are reduced sensitivity and dexterity of the fingers, vibration "white finger" as well as muscle, joint and bone disorders. Vibration "white finger" (also called "dead finger") is caused by hand–arm vibration. The main symptom of this disorder is a reduction of blood flow in the fingers leading to discoloration of the skin. The fingers feel cold and become numb, which in time can actually lead to necrosis of the fingertips. The condition is aggravated by cold.

The most common frequency range for handheld motorized tools is between 25 and 150 Hz. In practice, most vibration consists of several separate vibrations at different frequencies and in different directions. From the individual characteristics of these vibrations, it is possible to calculate an average measure of the vibration level. This average level can then be used in practice to assess the impact of the vibration.

4.2.1   Guidelines on Vibration

This section contains guidelines on whole-body vibration and hand–arm vibration as well as shocks and jolts.

Avoid heath and safety risks from vibration

To prevent health and safety risks from vibration, it is recommended that vehicles, machines and such be designed to keep the vibration level at any time below 0.5 m/s for whole-body vibration and below 2.5 m/s for hand–arm vibration.

Prevent shocks and jolts

Shocks and jolts often arise together with vibration. Shocks and jolts with peak intensify more than three times higher than the average vibration level will increase the total vibration stress and should be avoided. The reader should note that the guidelines for vibration given here assume that no shocks or jolts occur.

4.2.2   PREVENTING VIBRATION

Vibration can be prevented at source, during the transmission between source and exposed person and, to a limited extent, at the individual level. In this section we deal with a few possible measures.

Tackle vibration at source

Large machines and motorized tools often constitute a source of vibration. Rotating movements generally cause less vibration than reciprocating movements, a fact worth remembering when designing or selecting machines and tools. Likewise, hydraulic and pneumatic transmissions are superior to mechanical transmission in this respect. Heavy machines (those with a large mass) also generally cause less vibration.

Maintain machines regularly

Machines and handheld tools sometimes display loose fits, eccentricity or imbalance, all of which cause vibration, noise and wear. Regular maintenance is therefore very important.

Prevent the transmission of vibration

Whenever measures at source are inadequate, attention should be devoted instead to reducing the transmission of vibration. This is best done by damping the vibration where it enters the body, for instance, by fitting floors, seats and handgrips with a damping material. An example is a well-damped seat in a bus, which makes it difficult for the vibrations to reach the body from the floor. The seating surface is fitted with a damping material and a pneumatic spring is located between the seat and floor for damping.

If necessary, direct the measures at the individual

If measures at source and in transmission are not effective, then attention must be directed at the individual. This can be done by reducing the duration of exposure, for example by alternating tasks which entail vibration with tasks that do not entail vibration. Cold, humidity and smoking increase the risk of vibration "white finger" and can be counteracted at the individual level, among other ways, by using gloves for protection against cold and humidity.

4.3   ILLUMINATION

Illumination can affect a person's performance and well-being. The light intensity, which is the amount of light that falls on the work surface, must be sufficiently high whenever visual tasks have to be carried out rapidly and with precision and ease. Apart from light intensity, the differences in luminance (contrast) in the visual field are also important. Luminance is the amount of light reflected back to the eyes from the surface of objects in the visual field.

The color of the light and the presence of daylight can affect a person's mood and therefore performance.

Light intensity is expressed in lux, and luminance (brightness) in candela per m^2 (cd/m^2). Color temperature is in Kelvin (K).

4.3.1   GUIDELINES ON LIGHT INTENSITY

In determining the amount of light that must fall from the surroundings onto a work surface, it is necessary to distinguish between orientation lighting, normal working lighting and special lighting.

Select a light intensity of 20–200 lux for orientation tasks

A light intensity of 20–200 lux is sufficient where the visual aspect is not critical, for example, in the corridors of public buildings or for general activities in store rooms, provided no reading is required. The minimum required intensity to detect obstacles is about 20 lux. A higher light intensity may be necessary for reading notice boards and the like, or to prevent excessive differences in brightness between adjoining areas; this allows the eyes to adjust more rapidly when moving between the areas, such as when driving into tunnels. Adjustment of the eye can take a fairly long time if the differences in brightness are large.

Select a light intensity of 200–750 lux for normal activities

Reading normal print, operating machines and carrying out assembly tasks can be considered normal visual tasks, and the following guidelines apply in this instance:

- A light intensity of 200 lux is adequate if the information is large enough and visual tasks are not critical such as in waiting rooms or archives.
- Greater light intensities are necessary if visual tasks are more important (e.g., reading and writing in offices) and if the details are small or if the contrast is poor.
- People with limited vision and older persons require more light.
- A greater light intensity is sometimes required to compensate for large differences in brightness between and within rooms, due, for example, to high light intensites in neighboring rooms or to the presence of windows.

Select a light intensity of 750–5000 lux for special applications

It is sometimes necessary to use localized task lighting. This can compensate for shadows or reflections on the work surface. For special activities such as visual inspection tasks, much higher illumination levels are used to enable fine details to be distinguished.

4.3.2 GUIDELINES ON BRIGHTNESS DIFFERENCES

This section provides some guidelines on differences in brightness within the visual field.

Avoid excessive differences in brightness in the visual field

Excessive differences in brightness between objects or surfaces in the visual field are undesirable. Large differences can result from reflections, dazzling lights and shadows, among other things. Table 4.3 shows a few examples of how people experience differences in brightness (expressed as the luminance ratio, which is the brightness of one object divided by the brightness of another).

Limit the brightness differences between the task area itself, the close surroundings and the wider surroundings

The visual field can be divided into three zones: the task area, the close surroundings and the wider surroundings. The brightness of the task area should not be three times larger or three times smaller than that of the close surroundings. The brightness of the task area should not differ from that of the wider surroundings by more than a factor of ten. Differences in brightness that are too small should also be avoided because this makes a room look boring.

TABLE 4.3
Perception by Humans of a Few Luminance Ratios

Luminance ratio	Perception
1	None
3	Moderate
10	High
30	Too high
100	Far too high
300	Extremely unpleasant

Luminance ratios greater than 10 are considered excessive.

Note: Excessive differences in brightness within the visual field must be avoided.

4.3.3   GUIDELINES FOR THE COLOR OF THE LIGHT

Daylight is white light and has a color temperature of 5000–6500 K. Light with low color temperature (<5000 K) has more yellow and red and is perceived as warm (similar to the glow of metal or fire). Light with high color temperature (>6500 K) has more blue and is perceived as cold.

Avoid too cold and too warm colors for indoor lighting

For indoor lighting too warm and too cold colors of the light must be avoided. Too warm color of the light may be too stimulating and too cold colors too boring. In working environments, most people prefer warmer colors of the light (3000–5000 K).

4.3.4   GUIDELINES FOR IMPROVING LIGHTING

Steps taken to improve lighting aim mainly to provide sufficient light intensity, and to avoid excessive brightness differences in the visual field such as may occur with light sources, windows, reflections and shadows.

Ensure good legibility of information

When the visibility of the information is insufficient, it is more effective to improve the legibility of the information than to increase the light intensity. Further increases in light intensity are pointless when lighting is already intense. The legibility of information can be improved by enlarging the details (e.g., by using a larger typeface or smaller reading distance) or by increasing the contrast (e.g., black letters on a white background). Recommendations for the presentation of information are given in Chapter 3.

Combine ambient and localized lighting

Except for orientation tasks, the required light intensity on a work surface can be achieved by a combination of fairly limited ambient lighting and more intense localized or task lighting. The desired ratio between the general and the localized light intensity is determined among other ways by the criteria on brightness difference between task and surroundings (see Guidelines on brightness differences, p. 77), and by personal preference. The intensity of any localized lighting must be adjustable.

Daylight can also be used for ambient lighting

Available daylight should also be used for general lighting. Incoming daylight and a view to the outside are much appreciated by most people. Using blinds can prevent large variations in daylight intensity from direct sunlight. Excessive brightness differences in the visual field (see p. 77) can occur in workplaces close to windows.

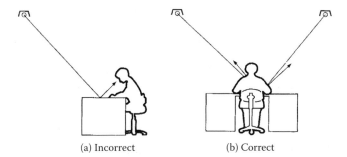

(a) Incorrect   (b) Correct

FIGURE 4.4   Light sources must be located in such a way that reflections and shadows are avoided.

Screen sources of direct light

Blinding by direct light can be avoided by screening off light that radiates sideways. However, vertical surfaces are then less well illuminated. This can be compensated for by opting for a light interior environment.

Prevent reflections and shadows

Light sources must be located relative to the workplace so as to prevent reflections and shadows. Figure 4.4 shows the optimum location of light sources for a workplace. In VDU workstations, special care is required to prevent reflections on the screen.

Use diffuse lighting

Excessive reflections can be avoided by using indirect (i.e., diffuse) lighting in ceilings. Table tops, walls and such should also produce diffuse reflection of any incident light; in other words, the reflected light should be evenly distributed in all directions. The surfaces should therefore have a matt finish. The optimum amount of light reflected from a surface (reflectance) depends on the purpose of that surface. Recommended reflectance values are given in Table 4.4. The reflectance is a value between zero and one, with a zero value meaning that no light is reflected (dark surface) and a value of one meaning that all the light is reflected (light surface).

TABLE 4.4
Recommended Values for the
Reflectance of Various Surfaces

Surface	Reflectance
Ceiling	0.80–0.90 ("light")
Walls	0.40–0.60
Table tops	0.25–0.45
Floor	0.20–0.40 ("dark")

Avoid flicker from fluorescent tubes

Fluorescent tubes produce a flickering light that can be disturbing. It is possible to avoid any detectable flicker if two or more tubes are placed in a mounting so that they alternate out of phase; this is achievable by correct connection to the grid. The use of a single fluorescent tube, for example, to reduce power consumption, is ill advised.

4.4 CLIMATE

The indoor climate needs to satisfy several conditions if work is to be carried out in comfort. Four climatic factors (air temperature, radiant temperature of surfaces, air velocity, and relative humidity) are significant in this respect. Whether a climate is considered pleasant depends also on the level of physical effort required by the work and on the type of clothing. Work is sometimes carried out in very cold environments such as in cold-rooms or outside, or in very warm environments such as near ovens. Special precautions are then necessary to prevent freezing or burning of the exposed skin, mainly on the face and hands. Without these precautions, the time spent in cold or hot environments has to be limited.

4.4.1 GUIDELINES ON THERMAL COMFORT

This section contains guidelines on the four climatic factors of air temperature, relative humidity, radiant temperature and air velocity.

Allow people to control the climate themselves

Whether people find a climate pleasant depends very much on the individual. The aim must therefore be to allow people to control the climatic factors as much as possible themselves. This is feasible, for example, in an office with separate rooms.

Adjust air temperature to the physical demands of the task

Table 4.5 contains global guidelines on air temperature for tasks requiring different levels of physical effort. The guidelines ensure that people will feel comfortably cool

TABLE 4.5

Guidelines on Air Temperature for Tasks Requiring Different Levels of Physical Effort

Type of Work	Air Temperature (°C)
Seated, thinking task	18–24
Seated, light manual work	16–22
Standing, light manual work	15–21
Standing, heavy manual work	14–20
Heavy work	13–19

to comfortably warm. The assumption here is that the relative humidity is 30 per cent to 70 per cent, that the air velocity is less than 0.1 m/s, and that normal clothing is worn.

Avoid too humid and too dry air

Humid air (relative humidity in excess of 70 per cent) or dry air (relative humidity less than 30 per cent) can affect thermal comfort. Dry air can lead to irritation of eyes and mucous membranes, and also increases the possibility of static electricity (risk of inflammation or ignition of chemical substances, unpleasant shocks, equipment failure). The humidity can be controlled either by adding moisture to the air or by removing it.

Avoid radiating surfaces

Hot surfaces such as a roof can affect thermal comfort. Steps must be taken whenever the radiant temperature of these surfaces differs by more than four degrees from the air temperature (see climate control, p. 82).

Prevent drafts

Drafts can affect thermal comfort, mainly in the case of light work. Drafts are uncomfortable at air velocities above 0.1 m/s. Drafts can be caused, among other things, by ventilation (see pp. 82, 86).

4.4.2 GUIDELINES ON HEAT AND COLD

Hot and cold environments are not only uncomfortable; hot climates, such as near ovens, can be energetically very stressful to the heart and lungs. In addition, burns or frost can injure parts of the body.

Avoid exposure to extremely hot or cold environments

Exposed parts of the skin can reach the threshold of pain in extremely hot climates or near very hot radiating surfaces. In a very cold environment, the hazard is frostbite, the risk of which increases at high air speeds.

Materials, which must be touched, should be neither too cold nor too hot

If bare skin comes into contact with very cold metal, it may adhere to the metal surface. To be on the safe side, the temperature of metals likely to be touched should be at least 5°C. A lower value can be tolerated for objects made of dry plastic or dry wood.

Table 4.6 shows the maximum temperature of materials allowed if skin burns due to contact are to be avoided.

TABLE 4.6

Maximum Temperature Allowed Depending on the Exposure Duration and the Type of Material

Duration of Contact	Type of Material	Maximum Temperature (°C)
Up to 1 min	Metals	50
	Glass, ceramics, concrete	55
	Plastics (perspex, Teflon™), wood	60
Up to 10 min	All materials	48
Up to 8 h	All materials	43

Note: Materials that have to be touched should not be too hot if burns are to be avoided.

4.4.3 CLIMATE CONTROL

In this section we discuss measures relating to thermal comfort, and to hot and cold climates.

Locate equally heavy tasks together in a room

It is desirable that tasks, which are more or less equally heavy, should be located together in a separately heated room. This makes it possible to achieve a pleasant climate for each group of tasks.

Adjust the physical demands of the task to the external climate

It is not possible to control the outdoor climate, but cold and hot outdoor climates, to a certain extent, can be tolerated better by adjusting the energy demand of the task. In a cold climate, tasks should be heavier so as to increase body temperature and reduce the risk of freezing. In a hot climate, the opposite applies.

Optimize air velocity

Where there is a draft (maximum 0.1 m/s), it is sensible to increase the air temperature to allow work to be carried out in comfort. In very cold climates the air velocity should always be as low as possible to prevent parts of the body from freezing. Conversely, a hot environment becomes more pleasant if the air velocity increases.

Prevent unwanted radiation

Insulating or screening off radiating surfaces such as walls, floors, roofs and windows can suppress radiation. In addition, correct layout of the work space can help increase the distance between the person and the source of radiation. Finally, the air temperature can also be adjusted to reduce the difference between air temperature and radiant temperature.

Limit the time spent in hot or cold environments

People themselves should be able to determine how much time they are able to spend in hot or cold environments.

Use special clothing when working for long periods in hot or cold environments

Clothing with a high insulation value affords protection against cold. Similarly, special clothing can also be used as protection against heat (e.g., for firemen).

4.5 CHEMICAL SUBSTANCES

Chemical substances occur in the environment as liquids, gases, vapors, dusts or solids. Some substances can cause discomfort or present a health hazard if inhaled or ingested or if they come into contact with the skin or eyes. The symptoms can develop immediately or at a later stage. It is known that many substances are irritants, carcinogens, mutagens (damage genes), or teratogens (lead to birth defects). The body must therefore be exposed as little as possible to such chemical substances.

4.5.1 GUIDELINES ON CHEMICAL SUBSTANCES

The most important guidelines on chemical substances, given in this section, are based on the so-called TLVs (threshold limit values). These are official international limits for chemical substances in air and are intended to prevent adverse health effects (rather than merely discomfort).

Apply TLVs or other limits as maxima for chemical substances in ambient air

TLVs are available for several hundred substances. The list of TLVs is regularly updated by the inclusion of new substances and by taking account of new data on the toxicity of substances. The TLV is an 8-hour weighted average concentration and should not be exceeded in any single day.

Certain substances have a rapid toxic effect, in which case a separate TLV is established, namely the TLV-C (C = ceiling), which may not be exceeded at any time (see below). Only a small proportion of the known chemical substances appear in the TLV list. Whenever a chemical substance does not appear in a national TLV list, the lists of other countries or toxicological handbooks can be consulted. If the substance appears in none of these, this still does not mean that it is harmless. In this event, individual organizations often tend to apply their own standards.

Avoid carcinogenic substances

Certain airborne substances are known to cause cancer. Exposure to these substances must be avoided at all times. Table 4.7 shows a random selection of chemical substances which could be present in air and are considered to induce cancer. A comprehensive overview of known or suspected carcinogens can be found in publications of the International Agency for Research on Cancer (IARC), Lyon, France.

TABLE 4.7
Random Selection of Chemical Substances That Are Considered to Induce Cancer

Substance	Example of Use
Asbestos	Thermal insulation
Benzene (benzol)	Solvent
Chrome compounds	Pigment
Polycyclic hydrocarbons	Component of tar
Vinylchloride	Raw material for PVC

Avoid peak exposures

Short-term exposure to high concentrations of a chemical substance can affect health even if the TLV is, on average, not exceeded over an 8-hour period. Therefore, the TLV-C values should be applied for substances with a rapid toxic effect. The design of a work or living environment must ensure that the TLV-C is not exceeded in other circumstances, such as cleaning or maintenance.

Exposure to mixtures of substances should be avoided

Although in practice one is confronted mainly with mixtures of substances, there are generally no TLVs for such situations. There is also no guarantee that satisfying the individual TLVs will avoid health risks, as the effects of the individual substances can reinforce each other.

Always aim to remain as far below the TLVs as possible

It is important to try to remain as far as possible below the TLVs at all times. The rule of thumb in the design of new work or living environments is to achieve concentrations of less than a fifth of the TLVS. Remember also that remaining below the TLVs does not guarantee the absence of any discomfort (e.g., pungent odors). Conversely, substances that cause no discomfort can in fact be dangerous.

Packages of chemicals should be labeled appropriately

The supplier of a chemical substance must provide information on the toxicity of the substance and how to use it. The first indication of this must appear on the label, which should bear appropriate standard warning signs (Figure 4.5).

4.5.2 MEASURES AT SOURCE

Measures can be aimed at the source or at the exposure. Measures at source are preferable, especially if this mean replacing the source. If this is not feasible, the source should be reduced. If this is still inadequate, then the source must be isolated. Measures taken at source can be directed at the chemical substance itself, the (production) process, or the working method.

Explosive	Flammable	Corrosive
Oxidizing	Toxic	Harmful or Irritating

FIGURE 4.5   Warning signs for chemical substances.

Remove the source

A fundamental measure at source is to replace the harmful substance by substances which, so far as is known, are not harmful or at least are less so. Examples of this are the use of water-soluble paints instead of solvent-based paints, and thermal insulation using rock wool rather than asbestos. Harmful production processes or working methods must be replaced by processes or working methods which are less harmful. An example of this is the use of an industrial vacuum cleaner instead of compressed air in cleaning activities.

Reduce the releases from the source

The reduction of emission at source can affect the chemical substance itself, the production process or the working method. Examples of measures aimed at the substance itself are the use of paint with a lower concentration of heavy metals, and the supply of raw materials in the form of a paste rather than a powder. Examples of measures aimed at the production process are reducing emissions by carefully tuning the process, carrying out regular maintenance, and reducing the fall height when emptying sacks of powder. An example of a measure aimed at the working method is to allow painted work pieces to dry in a separate room instead of in the spray cabinet.

Isolate the source of chemicals

A third measure aimed at the source is to prevent harmful substances from being released. An example is the use of enclosed, instead of open, transport systems for material, thus preventing the release of harmful chemicals.

4.5.3 VENTILATION

Measures aimed at the exposure route should be taken whenever those directed at the source are inadequate. In this section we discuss measures directed at the transfer between the source and people (ventilation of the air). The section that follows addresses measures aimed at individual exposure, such as organizational measures, or use of personal protective equipment.

Chemical substances must be extracted directly at source

If it is not possible to prevent the release of chemical substances, then harmful substances should be extracted directly at source. In such instances, the extracted air is often released to the environment without being cleaned. However, environmental laws restrict the permissible concentration of chemical substances in the exhaust air, and as these laws become more stringent, this will stimulate measures aimed at the source. Note that if harmful substances in air can be partly removed through an exhaust system, at the same time fresh air must be supplied to the workplace. Figure 4.6 shows an example of an exhaust system used in spray cabinets.

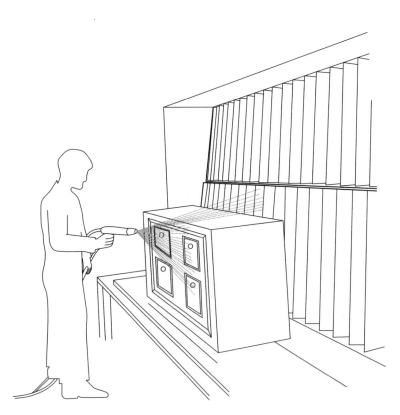

FIGURE 4.6   Use of an exhaust system in spraying activities, which removes residual paints and solvents from the air inhaled by the worker.

TABLE 4.8
Recommended Optimum Space and Air Change

Nature of the Work	Volume per Person (m³)	Fresh Air Supply Rates (m³ hr⁻¹)
Very light	10	30
Light	12	35
Moderate	15	50
Heavy	18	60

Provide an efficient exhaust system

In practice the effect of exhaust systems is often unsatisfactory because of less than ideal conditions, such as unfavorable location, inadequate maintenance, or faults. Avoid extracting the air only from higher up in the room, rather than from the breathing zone. Regular maintenance is essential to prevent dirt from reducing the efficiency of the system.

Pay attention to the effect on climate when designing air extraction and ventilation

Air extraction and ventilation increase the chance of a draft. This in turn influences the degree of thermal comfort (p. 80). It is also important that the fresh air supply be preheated.

Provide sufficient air changes

Indoor environments must also be adequately ventilated even if no dangerous substances are present. The required volume of fresh air per person and the rate of air change depend on the degree to which the work is physically demanding (Table 4.8).

4.5.4 Measures at the Individual Level

Measures to reduce the effect of chemical substances at the individual level are either organizational, whereby persons are exposed for as little time as possible to the substances, or else involve the use of personal protective equipment.

Implement organizational measures

Various organizational measures are possible to reduce people's exposure. People should spend as little time as possible in rooms with contaminated air, and likewise, the number of people exposed to the contaminated air should be limited. Activities where chemical substances are released can, for example, be separated from unaffected activities by locating them in a separate room. It is also feasible to carry out these activities outside normal working hours. The advantage is that fewer people are exposed to the substances. However, other precautions must be taken to protect those who are nevertheless exposed.

Use personal protective equipment

If measures aimed at the source or the exposure are not feasible or adequate, then personal protective equipment must be used, even though most users consider such equipment to be a nuisance.

In emergencies it is possible to use special masks fitted with filters, as these can provide protection against a number of gases. Special masks can also be used against fine dust. The masks must closely fit the shape of the face and make proper contact with the skin. Instruction in the use of masks is essential.

Use dust masks only for protection against coarse dust

Dust masks offer no protection against gases. They can screen part of the coarse, chemically harmless dust particles, but are inadequate at high concentrations (for example, if a mist hangs in the room).

Use protective equipment

Adequate protective equipment, such as gloves and aprons, should be worn when working with liquids that can be absorbed through the skin. Gloves are often considered to be a nuisance, but the protective effect of special skin creams is unproven.

Ensure a high standard of personal hygiene

Other kinds of measures can be taken to reduce the absorption of chemical substances through the skin:

- Clean dirty clothing and gloves regularly.
- Do not use dirty cleaning cloths.
- Cleanse the skin regularly with soap and water.
- Ensure rapid treatment of skin lesions.

SUMMARY CHECKLIST

NOISE

Guidelines on Noise

1. Is the noise level below 80 decibels?
2. Is annoyance due to noise avoided?
3. Are rooms perhaps too quiet?

Noise Reduction at Source

4. Has a low-noise working method been chosen?
5. Are quiet machines used?
6. Are machines regularly maintained?
7. Are noisy machines enclosed?

Noise Reduction through Workplace Design and Work Organization

8. Is noisy work separated from quiet work?
9. Is there an adequate distance from the source of noise?
10. Is the ceiling used for noise absorption?
11. Are acoustic screens used?

Hearing Conservation

12. Are hearing conservation measures suited both to the noise and to the user?

VIBRATION

Guidelines on Vibration

13. Are health and safety risks from vibration avoided?
14. Are shocks and jolts prevented?

Preventing Vibration

15. Is vibration tackled at source?
16. Are machines regularly maintained?
17. Is the transmission of vibration prevented?
18. Are measures at the individual level applied only as a last resort?

ILLUMINATION

Guidelines on Light Intensity

19. Is the light intensity for orientation tasks in the range of 20–200 lux?
20. Is the light intensity for normal activities in the range of 250–750 lux?
21. Is the light intensity for special applications in the range of 750–5000 lux?

Guidelines on Brightness Differences

22. Are large brightness differences in the visual field avoided?
23. Are the brightness differences between task area, close surroundings and wider surroundings limited?

Guidelines for the Color of the Light

24. Are too cold and too warm colors for indoor lighting avoided?

Improved Lighting

25. Is the information easily legible?
26. Is ambient lighting combined with localized lighting?
27. Is daylight also used for ambient lighting?
28. Are direct light sources properly screened?
29. Are reflections and shadows avoided?
30. Is diffuse light used?
31. Is flicker from fluorescent tubes avoided?

CLIMATE

Guidelines on Thermal Comfort

32. Are people able to control the climate themselves?
33. Is the air temperature suited to the physical demands of the task?
34. Is the air prevented from becoming either too dry or too humid?
35. Are radiating surfaces avoided?
36. Are drafts prevented?

Guidelines on Heat and Cold

37. Is the exposure to extreme hot or cold environments avoided?
38. Are materials that have to be touched neither too cold nor too hot?

Climate Control

39. Are equally heavy tasks grouped in the same room?
40. Are the physical demands of the task adjusted to the external climate?
41. Is the air velocity optimized?
42. Is undesirable radiation prevented?
43. Is the time spent in hot or cold environments limited?
44. Is special clothing used when spending long periods in hot or cold environments?

CHEMICAL SUBSTANCES

Guidelines on Chemical Substances

45. Is the concentration of chemical substances in air subject to limits (TLVs or other)?
46. Are carcinogenic substances avoided?
47. Are peak exposures prevented?
48. Is exposure to mixtures of substances avoided?
49. Is the concentration of chemical substances as far below the TLVs as possible?
50. Are packages of chemicals appropriately labeled?

Measures at Source

51. Can the source be removed?
52. Can releases from the source be reduced?
53. Can the source be isolated?

Ventilation

54. Are chemical substances extracted directly at the source?
55. Is the air exhaust system efficient?
56. Has attention been paid to climate at workplaces where exhaust and ventilation are used?
57. Are sufficient air changes provided?

Measures at the Individual Level

58. Are organizational measures possible?
59. Is personal protective equipment available?
60. Are dust masks used only as protection against coarse dust?
61. Are protective clothing and gloves available?
62. Is attention paid to personal hygiene?

5 Work Organization Jobs and Tasks

Activities of human beings usually take place in a wider organizational context. The activities of one person are related to the activities of others. Usually an organization is divided into units. An example is a post office. This office has different organizational units, for instance, concerning sales, mail delivery and financial matters. Within a unit people are employed in certain jobs. Let's consider a job at the stamp window. The employee has different tasks such as selling stamps and paying money orders. To perform the task to sell stamps, different actions are needed, such as handling the stamps, accepting the money and returning the change. The relations between jobs, tasks and actions are presented in Figure 5.1.

In this chapter recommendations are given for the design of work organization, jobs and tasks. The focus is primarily on paid work. However, most of the guidelines are also applicable outside of paid work (household, sport, voluntary work, traffic).

5.1 TASKS

The design of work organizations, jobs and tasks always begins with a description of all the tasks to be performed.

Describe tasks in a neutral way and not the way they are performed

There are different ways to perform a task. The task "selling stamps" can be done by a human being as well as by a machine. Both humans and machines can perform the tasks in different ways (for instance, the procedure to accept money or the type of machine involved).

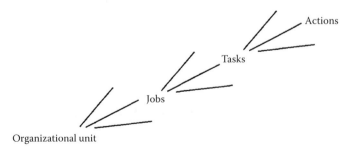

FIGURE 5.1   Jobs consist of tasks that, in turn, consist of actions.

When dividing between humans and machines, keep in mind their specific qualities

After the tasks have been determined they have to be distributed between humans and machines. This process is called allocation of tasks. Some tasks can be better performed by humans, others by machines. The following guidelines can be used in deciding when to allocate tasks to people rather than machines.

- People are more creative than machines in solving problems, especially unexpected ones.
- People can communicate by several means, such as speech, visual expressions or gestures.
- People would rather not allocate certain tasks to machines, for instance, weighing alternatives in order to make decisions.
- People are sometimes cheaper than machines. This is particularly true for complex movement patterns that are required only occasionally.
- People are better at filtering relevant information from a mass of information.
- Machines are better at counting and repetitive actions.
- Machines can operate in unhealthy and extreme conditions.

Economic factors often play a crucial role in deciding the allocation of a task which is otherwise equally suited to people or machines.

5.2 JOBS

After the allocation of tasks, the tasks have to be combined into interesting jobs for people. The following criteria must be met if a job is to be interesting:

- Completeness of the job
- Control over the work
- Absence of repetitive tasks
- Alternation of difficult and simple tasks
- Freedom of workers to determine the method, the sequence of operations and pace within the job
- Opportunity to make contact with others
- Availability of information

These attributes are discussed in detail below.

Jobs must consist of more than one task

A job is complete if it consists of a logical, coherent group of preparatory, production and support tasks.

Opportunities to learn and control will arise mainly through a job's support and preparatory tasks. The terms *job enrichment* or *vertical job enlargement* are used to describe the process of making a job more complete. By contrast, assigning more of the same type of tasks to a job is called *job extension* or *horizontal job enlargement.*

Everyone contributes to solving problems

Helping to solve problems makes a job interesting. This is particularly true if the problems go beyond run-of-the-mill difficulties to include the unexpected. Solutions are usually arrived at through consultation, which can be of the following nature:

- Functional consultation: maintaining contacts with immediate colleagues;
- Work consultation: regular consultation with colleagues and line management;
- Working parties: the solution of particular problems by separately formed groups.

A prerequisite for all these forms of consultation is that not only should discussion take place, but also agreement on solutions must be reached; furthermore, the problems should be such that solutions do exist.

The cycle time must exceed one and a half minutes

Many jobs require repeating the same type of actions. This is most clearly illustrated by conveyor belt work. The time between two repeats, the cycle time, should not be too short, preferably not less than one and a half minutes. Shorter tasks are totally mind-numbing and should not constitute the main component of a job. The ability to influence the task is largely absent in short cycle tasks: the work pace and the location, as well as the starting and finishing times, are all fixed.

Alternate easy and difficult tasks

A job should be made up of both simple and difficult tasks. If a job consists entirely of difficult tasks, there is a risk of mental overstress. If there are too many easy tasks, the worker will not feel challenged and boredom will set in.

Allow people to decide independently how to do their work

People will find a job more interesting if they can decide independently how to do their work. This autonomy can relate to the method of working, the order in which the individual actions are carried out, and the place of work. Other aspects are the possibility of rejecting supplied material (e.g., substandard raw materials, components, information) and the authority to call on the help of others.

Provide contacts

Contacts with others must form a component of the job. A contact can take place in several ways:

- By helping one another
- By discussing the work
- By talking of things other than work

The job can be made more interesting by widening the diversity of contacts or extending the time spent making these contacts. Informal contacts take place when the employee leaves his or her workplace once in a while or when employees work in close proximity. A noisy environment adversely affects opportunities for contact.

Tasks should be accompanied by sufficient information

A sustained flow of information is needed if the best use is to be made of the opportunities people have of properly controlling their tasks. This implies information at two levels, namely the workplace level and the division or company level.

The type of information can also be of a dual nature:

- Feedback: people subsequently receive information on the quality and quantity of work produced
- Forward coupling: prior information on the required quality and quantity, as well as on the influence of the pace of work on these two factors

This information must help highlight the objective and the results of the work.

5.3   WORK ORGANIZATION

When combining jobs in a work organization the following aspects are relevant:

- Flexible forms of organizations
- Autonomous groups
- Coaching management styles

These aspects will be described below. A special focus is put on the design of autonomous groups and the new role of the management to support the optimal performance of the employees within the new developments.

5.3.1   FLEXIBLE FORMS OF ORGANIZATIONS

Organizations must be able to react quickly to changing environments and an ongoing process of renewal of products, services and labor processes.

Replace hierarchical work organizations with more flexible structures

The traditional hierarchical work organization is increasingly being replaced by more flexible structures (Figure 5.2). The layers of the organization disappear (more flat organization) and the boundaries between the organizational units become blurred (more cooperation to serve the customers). More tasks and responsibilities

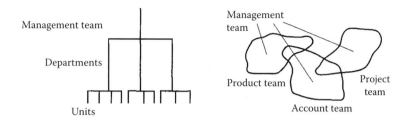

FIGURE 5.2 Hierarchical structures are being replaced by more flexible forms.

are allocated to employees lower in the organization. Workers work together in self-supporting teams to use one another's strengths. The role of the manager changes from boss to coach.

To change an organization in this way, bravery and persistence are needed from the management. It is attended by a change of culture: another way of communication, control and justification. Extra attention is needed for education, training and implementation. External support may be needed to guide these processes.

Make housing conditions flexible

A flexible form of organization means that employees are judged less on their presence at work, but more on the result of their activities. This means that workers should not always have their own individual offices (or rooms) but that the space is used more flexibly. Sometimes it is possible to perform part of the work activity at home instead of at the workplace. In an office environment this could mean that workers take the first free space when they arrive. Their personal belongings can accompany them in a movable drawer.

Make working times flexible

A flexible form of organization also means that working and resting hours can be taken more flexibly and that they can be determined by the social environment or by the traffic conditions. Of course, this form of organization is not applicable to everyone; in industry the occupation of machines and the dependency of suppliers play important roles.

5.3.2 AUTONOMOUS GROUPS

Teamwork is an alternative for the fixed forms of organization with individual control of employees. An autonomous group is a fixed group of employees who, together, are responsible for the total process in which products or services are realized without continuously consulting a manager.

This form of organization can contribute to shorter production times, higher productivity, higher quality, more innovations, higher flexibility, improvement of the quality of work and better labor relations. Above all production risks are smaller because people are in more than one position in the team. Some guidelines for the operation of groups follow.

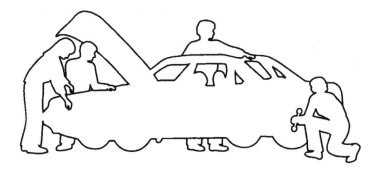

FIGURE 5.3   The group must work toward a recognizable goal.

The assignment to the group must be clear

The assignment to the group must consist of strongly related activities. The assignment must be clear and the result must be identifiable and measurable. The starting situation is described precisely and the members of the team depend on each other. The team has sufficient possibilities to make decisions: decide within the conditions how the product is realized. An example is the cooperative work on a car (see Figure 5.3).

The team size must be 7–12 members

The optimal size of teams is 7–12 members. The size is determined by

- The involvement of team members
- The time span in which decisions can be made
- The productivity
- The ability of the group to solve problems

In larger teams, the productivity and ability to solve problems may be better and the other aspects may be worse. Smaller teams need less time to decide and have more involvement.

5.3.3.   COACHING MANAGEMENT STYLES

In a flexible organization where the responsibilities are lower in the organization, the role of the management changes.

Don't act as the boss, but coach the employees

In the new management style that fits flexible organizations better, the manager is no longer the boss but acts as coach. The differences between these styles of management are presented in Table 5.1.

TABLE 5.1
The Difference between Boss and Coach

The Boss
- The boss does not tell everything, he keeps some information back.
- The boss knows everything; you can ask him everything.
- The boss knows everything better and clearly shows this to be so.
- The boss solves the problems.
- The boss is always busy with his own work.
- The communication between boss and employee is restricted to work instructions.
- If you work well you hear nothing; when something goes wrong you are in trouble.
- When things go wrong, it is always a mistake of the employee.

The Coach
- The coach presents information to those involved as soon as possible.
- The coach does not need to know everything.
- The coach and the employee depend on each other.
- The coach facilitates employees to solve their own problems.
- The coach supports the work of the employees.
- The communication between coach and employee is two-way.
- The coach is interested in the employee; he asks questions and listens.
- When things go wrong, the coach asks himself what to do to prevent mistakes.

SUMMARY CHECKLIST

TASKS

1. Are tasks described in a neutral way?
2. Has a conscious decision been made about allocating tasks to a person or to a machine?

JOBS

3. Does the job consist of more than one task?
4. Do those involved contribute to problem solving?
5. Is the cycle time longer than one and a half minutes?
6. Is there alternating between easy and difficult tasks?
7. Can those involved decide independently on how the tasks are carried out?
8. Are there adequate possibilities for contact with others?
9. Is the information provided sufficient to control the task?

WORK ORGANIZATION

Flexible Forms of Organization

10. Are hierarchical work organizations replaced by more flexible structures?
11. Are housing conditions flexible?

12. Are working times flexible?

Autonomous Groups

13. Is the assignment to the group clear?
14. Does the group consist of 7–12 members?

Coaching Management Styles

15. Is the role of the manager more coach than boss?

6 The Ergonomic Approach

An ergonomic approach can be adopted in virtually any kind of design or purchasing project. Such an approach merely requires the systematic application of ergonomic principles. The person providing the ergonomic input to the project, whom we will refer to as the ergonomist, must work systematically and, wherever necessary, should call upon other specialists. In this chapter we present an appropriate general methodology for such an approach, which might be applicable to the following types of projects:

- Selecting a commercially available product for purchase
- Improving an existing product or system
- Designing a new product or system
- Adapting an individual workplace
- Refurbishing a business or workplace, for instance, after automation
- Designing a complete plant

These guidelines do not provide a universal methodology because of the wide variety of possible projects. Thus, for many projects some of the guidelines will be omitted because they are not applicable. The recommendations given here are addressed to the ergonomist, but when he is part of a team, they apply also to the other team members. The specific roles that the ergonomist can fulfill are those of:

- Subject specialist and
- Intermediary between technical designers and users and management

The chapter concludes with a control list, or so-called checklist, which can be used to ascertain that no ergonomic aspects of importance to the project have been neglected. The checklist is based on the knowledge gained in the previous chapters.

6.1 PROJECT MANAGEMENT

Involve users in the project

An important characteristic of the ergonomic approach is the involvement of users and other stakeholders in a project, in the earliest possible phase ("participatory ergonomics"). The goal of this policy is:

- Avoidance of mistakes in design or purchase
- Enlargement of acceptation
- Development of more ideas

- Faster identification of bottlenecks
- Giving workers a say in their work and enlarging both autonomy and human well-being

This procedure is especially important when many people are going to use the design or product, if a wrong design has serious consequences, or when there is mistrust and rejection.

The method is less applicable when the project has to be finished fast, when management is not open to the principle of participation, or when there are no significant advantages (for instance, a solution without any human workers).

For a participatory ergonomics approach, guidance of an ergonomist is desirable to bring in ergonomics knowledge, to promote stakeholders' cooperation and to achieve concrete results.

Introduce ergonomic requirements as early as possible in the project

It is better to apply ergonomics from the outset (prevention) rather than retrospectively (cure). Ergonomic requirements should therefore be introduced as early as possible in a project and must play a role in every one of its phases. It is still all too often the case that ergonomists are only involved when a project or system is almost completed. They are then expected to append a sprinkling of ergonomics, which means that a fundamental contribution is out of the question.

Use conventional methods for project management

In order to contribute effectively to a project, the manner in which the ergonomist provides an input must be suited to the method that is usual for the project team. Design or purchasing usually involves the separate, successive stages (Figure 6.1).

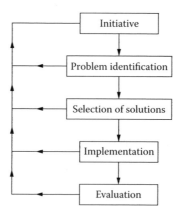

FIGURE 6.1   The successive stages of a design or purchasing project.

- Initiative: making a detailed survey of the project, formulating the questions and planning the rest of the project
- Problem identification: gathering the required data, which usually leads to a multitude of alternative solutions
- Selection of solutions: selecting from the alternatives and further developing the selected option
- Implementation: implementing the selected option
- Evaluation: evaluating the option and the project

These stages are described in more detail in the next five sections.

Make sure that the planning is flexible

The role of planning in any project is only to provide a logical sequence and to avoid aspects being overlooked. In practice there will always be an iterative process where any phase can lead to a revision of the previous one. For example, if in gathering data it appears that the project is still not properly defined, then a reformulation of the project becomes necessary.

Realize that cooperation implies joint responsibility

All those who work on a project share the responsibility for the final result. During a project an ergonomist must regularly ask himself whether it is worthwhile to continue contributing. It can happen that the ergonomist has to compromise to such an extent that the label "ergonomically designed" is not really applicable to the end product.

6.1.1 INITIATIVE PHASE

A project starts with a given objective, but this is seldom formulated in such a way that ergonomists can be involved immediately. Therefore we must attempt to describe the objective clearly and in a detailed manner. This requires a series of discussions with all parties.

Define who is involved in the project

It is important to the success of a project to know exactly who is involved in a project and who will have to deal with the result. The parties are:

- Clients (e.g., company managers)
- Members of the project team
- Help brought into the project, for example, subcontractors and experts
- Users and their environment

As soon as the names are known, it must be made clear how consultation will take place and how binding certain opinions or advice will be. The future users are best involved through a steering committee. The information on the project must be understandable to every person involved.

Make sure that those involved support the project

The support of all those involved in a project is essential to its success. In a business, this includes managers and workers, as well as suppliers. Their support also has a bearing on the course of the project and its outcome. Enthusiasm on their part should be cultivated by a good presentation at the beginning of the project, describing its course and potential benefits.

Agree on a code of conduct

It must be clear at the outset what freedom the ergonomist and other team members have in the project, such as access to rooms and documents. Agreement must also be reached on how the users consulted by the ergonomist or involved in a survey are informed of the result. This affects, for example:

- Access to documents and/or parts of the company and
- Possibilities and limitations for publicizing the future results

Ergonomists must always treat information about individuals in confidence. Indeed, even the client is given no access to the survey data on individuals.

Do not raise false expectations on the part of the users

Whenever an ergonomist approaches users in a project, the contact with them must be by prior arrangement. It is also worth bearing in mind that people, who are dissatisfied, sometimes expect the visit of an ergonomist to lead to immediate improvements. What is more, users sometimes associate such a visit with purely coincidental reorganization measures, such as dismissals or transfers. Care is certainly of the greatest importance when the aim of a visit is to obtain comparative data for a project from a similar part of the company.

State the limits of the project

It is mainly in production systems that a change in a section (e.g., a division or department) will have an influence on other sections. The boundaries of the project must therefore be properly established; these depend on available finance and time. In addition, management plans for the future must be made known.

Describe the planned course of the project at the outset

The intended course of the project must be absolutely clear from the start. Agreements about the evaluation phase must also be made at the beginning: when, how and with whom the evaluation will take place. Possible shortcomings in the solution can be identified through this evaluation, which usually also provides a retrospective view of the adopted working method. The latter can be useful in future projects of a similar nature.

6.1.2 PROBLEM IDENTIFICATION PHASE

The problem identification phase is aimed at gathering the relevant ergonomic aspects of the project on paper. It involves compiling complaints, ideas and wishes, and subsequently analyzing the ergonomic aspects in the light of

- Safety, health and comfort associated with the adopted solution and
- Performance, in other words, the usability of the adopted solution

The usability is determined both by objective and subjective elements:

- Objective: the efficiency (can the user work rapidly and without error with the product or system?)
- Subjective: the acceptance (does the user wish to work with the product or system and does its use not lead to stress?)

Use a checklist to avoid overlooking any aspects (see p. 111), and ensure that the aspects are examined and assessed not only in isolation but also in the context of the whole project.

Establish at the start how the data will be processed

Prior to gathering data it is necessary to reflect on the manner of its subsequent processing. The method of data processing must be tested beforehand if large amounts of data need to be compiled. It is also sensible at this point to prepare for a possible evaluation survey, which may involve, for example, the same persons having to be approached again.

Select more than one analysis technique

Various techniques can be used to gain an insight into the ergonomic aspects of the project; selecting the technique depends on the specific circumstances. It is generally sensible to use different techniques side by side in order to obtain a complete and reliable picture. Here are a few techniques:

- Analysis of documents and statistics, such as user statistics, absenteeism data, and registered complaints to provide a first impression of the project
- Observations of relevant events such as tasks and operations
- Interviews to provide an impression of the users' experience with problems through a questionnaire which is more or less structured, depending on circumstances
- Group discussions where the problem is discussed with a limited group of users (say, 6–12 people)
- Written questionnaire by which data are gathered from large groups of users

- Experimental methods by which some ergonomic aspects are investigated in a controlled manner in a laboratory or in the field and the data are obtained by measurements on people or on their physical environment

The specific choice and effect of these techniques depends on the type of project.

Always start with a survey of existing documents

There is no point in investigating aspects that are sufficiently well described in existing documents. Therefore the ergonomic contribution should always start with a survey of all existing relevant information, which will include both project documentation and specialist ergonomic literature. Good overviews of the more recent general ergonomics literature are available (see also Chapter 7). Congress proceedings and dissertations also provide much information in addition to that contained in standards, books and journals, but are more difficult to trace.

Project documentation is important because the history of the project is often a determining factor in its future.

Make sure that the analysis does not influence the result

It sometimes happens that the act of observation or measurement influences the result. If people feel, or know, that they are under close scrutiny, their bodies will function differently (blood pressure, heart beat) and their behavior will change (anticipation of desired outcome). These influences must be kept to a minimum by allowing those involved to become fully accustomed to the analysis method.

6.1.3   SELECTION OF SOLUTIONS PHASE

Enough must be known about a project before possible solutions can be devised. On the other hand, the search for a solution should initially not be limited by what might appear to be rigid project constraints. When making a purchase we will, in this phase, assess all available products. In design processes we will draw up an inventory of options. This is a point where ergonomics textbooks, software and other tools can make an important contribution in formulating ideas (see Chapter 7).

Realize that textbooks, software and other tools
do not provide a complete answer

Textbooks, software (Figure 6.2) and other tools such as templates (Figure 6.3) and mannequins (Figure 6.4) never give a complete answer to questions such as how to design a control panel, a room or product.

Allow users to work with a prototype

Many suggestions for improvement are made by users who are allowed to see a prototype and are given the opportunity to use it. This will increase the acceptance of the system at subsequent stages and ensure the commitment of the users to the

FIGURE 6.2 Example of a software tool to analyze human lifting.

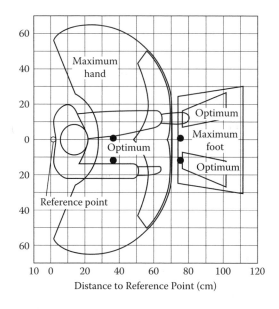

FIGURE 6.3 Example of a template that provides help in assessing drawings (view from above).

system. Mock-ups can be used for workplace layouts, that is, full-scale models of the design. The material from which these mock-ups are made must allow rapid modifications; for example, wood or cardboard are quite suitable.

Computer simulations can also be used to predict reactions and are most useful if they can later be used for training.

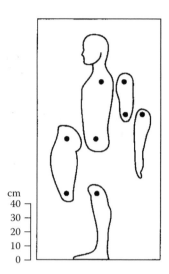

FIGURE 6.4   Example of a mannequin to illustrate the possibilities and limitations of the human body.

Remember the indirect users

Users other than direct users will also need to be able to deal with the new situation. Two such important target groups are cleaners and maintenance workers. Some designers make the mistake of assuming that cleaners and maintenance workers do not make full-time use of the design. However, this should not be taken to mean that suboptimal solutions are acceptable. Indeed, the working day of a cleaner or maintenance worker might otherwise always consist of a succession of unsatisfactory situations.

6.1.4   IMPLEMENTATION PHASE

The next phase is to put in place the product or system that is selected as the best alternative. The following aspects play a role here:

- Installation and testing
- Workplace design
- Training and support to users
- Organizational changes
- Acceptance of the new product or system

Each of these aspects is dealt with below.

Select an implementation strategy beforehand

There are various ways of achieving a smooth transition from an old product or system to a new one. The choice depends on the type of project:

- From scratch: There is no old system. Staff recruitment requires special attention (group processes, training).
- Direct transition: At a given moment the old product or system is completely replaced by a new one. Both the organization and the technology change simultaneously. The output immediately following transition will be lower.
- Parallel application: The old and new systems are used side by side. The users must be well informed about the advantages of the new system, otherwise they will tend to use the old system too much since, after all, they are familiar with it.
- Phased introduction: Successive phases of the new product or system are brought into use gradually. Acceptance can be more difficult because the advantage of the new solution is not immediately apparent to the users. Providing information is then very important. A particular form of this method is the introduction of a new system on a departmental basis.
- Occasional trial runs: This method is difficult to assess because the extra attention, which the new system receives, influences its use, Just as in the case of the analysis.

Make sure tests are realistic

When testing the system or products, the real situation should be reproduced as closely as possible. This means, for example, that an automated database must react in the dialogue just as if it were completely filled with data. Involving the users in the tests can also double as training.

Train all users

Everyone who is involved with the system must receive some form of training. The training does not necessarily have to be the same for everyone. It is conceivable that key personnel could be given extensive training, after which they in turn can train their colleagues.

Support the implementation by providing good manuals

Good written explanations are an essential part of the implementation of the system. After all, users cannot be expected to remember all the finer points of a verbal explanation. The written documentation cannot, however, replace the verbal explanation. What is more, the objections and requirements given in Chapter 3, such as loss and limited accessibility of documents, also apply. Manuals act rather as a form of memory support.

The documentation must be suited to the desired objective. If the manual is intended to be used merely for reference, there is no point in giving a chronological process description. However, many users need information when something goes wrong, so a manual must anticipate this. Many manuals describe operator actions, followed by their effect, whereas the user needs an overview of the effects, followed by the actions required. Manuals are definitely not a suitable means of compensating for compromises in the design.

Give the users a role in organizational changes

The tasks required to move from the old to the new organization must be properly assigned: responsibilities for specific aspects of the transition must be clear. These responsibilities are usually given to technical designers, instructors or ergonomists. It is also recommended to allocate roles to potential users, as this increases their commitment.

Convince the users of the improvements

If a project has led to a number of ergonomic improvements, it is crucial that the users be convinced of this. Some improvements, such as the disappearance of aches and pains, have an effect only in the long term. Any disadvantages must also be pointed out. Indeed, there is no sense in pointing out the advantages while glossing over the disadvantages. Management must also be convinced of the improvements offered by the new product or system.

6.1.5   EVALUATION PHASE

In time, it may appear, even in cases of careful selection or design, that all is not optimal. This will usually concern details where there is room for minor adjustments. Although many problems may appear to be solved in practice, the project still needs to be evaluated systematically after its introduction. It is necessary to assess whether the outcome meets the initial objectives or, to put it another way, whether the end result satisfies the initial objectives.

Keep to the same techniques of data collection

Whenever there is a redesign or readjustment, the result must afterwards be compared with the old situation. For this, it is advisable to use the same techniques as in the initial data collection, so that a proper comparison is possible. This means in the first instance a re-examination of the formal description of the activities on paper and subsequently forming a clear picture of the situation by means of observations, interviews, etc.

Allow teething problems to sort themselves out

It is important not to carry out an evaluation too soon after implementation. Initially there will inevitably be teething problems that could lead to incorrect assessment of a new product or production system. This is also true of the attitude of users toward the new situation. If the modification consists of automating a system, users will initially find it interesting to work with the new product. In the course of time, the novelty will wear off and working with the new system may sometimes appear more monotonous than working with the old one.

Beware of habit formation

People are adaptable, and this can be an advantage. However, their ability to adapt may distort the results of an evaluation because they may have adapted to an undesirable situation. We must, therefore, carefully examine whether the use of the new product or system takes place in the correct (i.e., intended), manner.

6.2   CHECKLIST

A checklist can be used in different phases of an ergonomic approach, for example:

- To avoid overlooking any aspects
- To detect where problems might develop
- To measure the effect of the implementation
- To obtain ideas for alternative solutions

Prepare your own checklist, based on that of others

In practice an ergonomist will use an existing checklist as a basis for developing his own specific list. A checklist for an office environment will look very different from a checklist for the steel industry, and will emphasize entirely different topics.

The checklist should sometimes be combined with a scoring list so as to be able to carry out a reasonably objective ergonomic assessment of situations and a comparison thereof. The checklist presented is aimed only at the main points and allows only "yes" and "no" answers. The topics for consideration are based on the recommendations in the previous chapters.

ERGONOMICS CHECKLIST

PROJECT FACTORS (CHAPTER 6)

GENERAL

1. Are users involved in the project?
2. Are ergonomic requirements introduced as early as possible in the project?
3. Are conventional methods used for project management?
4. Is the planning flexible?
5. Is it understood that cooperation implies responsibility?

INITIATIVE PHASE

6. Has who is involved in the project been defined?
7. Is the project supported by those who are involved?
8. Do team members agree on the code of conduct?
9. Are false expectations on the part of the user avoided?

10. Are the limits of the project stated?
11. Is the course of the project described at the onset?

PROBLEM IDENTIFICATION PHASE

12. Is the data processing established at the start?
13. Is more than one analysis technique selected?
14. Are existing documents surveyed at the start?
15. Is it assured that the analysis does not influence the result?

SELECTION OF SOLUTIONS PHASE

16. Is it understood that textbooks, software and other tools do not provide the complete answer?
17. Are users allowed to work with a prototype?
18. Are indirect users taken into account?

IMPLEMENTATION PHASE

19. Is an implementation strategy selected beforehand?
20. Are tests realistic?
21. Are all users trained?
22. Is the implementation supported by the provision of good manuals?
23. Do users have a role in organizational changes?
24. Are users convinced of the improvements?

EVALUATION PHASE

25. Are the same techniques of data collection maintained?
26. Can teething problems sort themselves out?
27. Is there awareness for the possibility of habit formation?

FACTORS RELATED TO WORK ORGANIZATION, JOBS AND TASKS (CHAPTER 5)

TASKS

28. Are tasks described in a neutral way?
29. Has a conscious decision been made about allocating tasks to a person, or to a machine?

JOBS

30. Does the job consist of more than one task?
31. Do those involved contribute to problem-solving?
32. Is the cycle time longer than one and a half minutes?

33. Is there alternating between easy and difficult tasks?
34. Can those involved decide independently on how the tasks are carried out?
35. Are there adequate possibilities for contacts with others?
36. Is the information provided sufficient to control the task?

WORK ORGANIZATION

Flexible forms of Organizations

37. Are hierarchical work organizations replaced by more flexible structures?
38. Are housing conditions flexible?
39. Are working times flexible?

Autonomous Groups

40. Is the assignment to the group clear?
41. Does the team consist of 7–12 members?

Coaching Management Styles

42. Is the role of the manager more coach than boss?

FACTORS RELATED TO POSTURE AND MOVEMENT (CHAPTER 2)

BIOMECHANICAL, PHYSIOLOGICAL AND ANTHROPOMETRIC FACTORS

Biomechanical Factors

43. Are the joints in a neutral position?
44. Is the work held close to the body?
45. Are forward-bending postures avoided?
46. Are twisted-trunk postures avoided?
47. Are sudden movements and forces avoided?
48. Is there a variation in postures and movements?
49. Is the duration of any continuous muscular effort limited?
50. Is muscle exhaustion avoided?
51. Are the breaks sufficiently short to allow them to be spread over the duration of the task?

Physiological Factors

52. Is the energy consumption for each task limited?
53. Is rest taken after heavy work?

Anthropometric Factors

54. Has account been taken of differences in body sizes?
55. Have the right anthropometric tables been used for specific populations?

Factors Related to Posture

56. Has a basic posture been selected that fits the job?

Sitting

57. Is sitting alternated with standing and walking?
58. Are the heights of the seat and backrest of the chair adjustable?
59. Is the number of adjustment possibilities limited?
60. Have good seating instructions been provided?
61. Are the specific chair characteristics dependent on the task?
62. Is the work height dependent on the task?
63. Do the heights of the work surface, the seat and the feet correspond?
64. Is a footrest used where the work height is fixed?
65. Are excessive reaches avoided?
66. Is there a sloping work surface for reading tasks?
67. Is there enough legroom?

Standing

68. Is standing alternated with sitting and walking?
69. Is the work height dependent on the task?
70. Is the height of the work table adjustable?
71. Has the use of platforms been avoided?
72. Is there enough room for the legs and feet?
73. Are excessive reaches avoided?
74. Is there a sloping work surface for reading tasks?

Change of Posture

75. Has an effort been made to provide a varied task package?
76. Have combined sit–stand workplaces been introduced?
77. Are sitting postures alternated?
78. Is a pedestal stool used once in a while in standing work?

Hand and Arm Postures

79. Has the right model of equipment been chosen?
80. Is the tool curved instead of the wrist being bent?
81. Are handheld tools not too heavy?
82. Are tools well maintained?
83. Has attention been paid to the shape of handgrips?
84. Has work above shoulder level been avoided?
85. Has work with the hands behind the body been avoided?

FACTORS RELATED TO MOVEMENT

Lifting

86. Have tasks involving manual displacement of loads been limited?
87. Have optimum lifting conditions been achieved?
88. Has care been taken that any one person always lifts less, and preferably much less, than 23 kg?
89. Have lifting situations been assessed using the NIOSH method?
90. Are the weights to be lifted not too light?
91. Are the workplaces suited to lifting activities?
92. Are handgrips fitted to the loads to be lifted?
93. Does the load have a favorable shape?
94. Have good lifting techniques been used?
95. Is more than one person involved in heavy lifting?
96. Are lifting accessories used?

Carrying

97. Is the weight of the load limited?
98. Is the load held as close to the body as possible?
99. Are good handgrips fitted?
100. Is the vertical dimension of the load limited?
101. Is carrying with one hand avoided?
102. Are transport accessories being used?

Pulling and Pushing

103 Are pulling and pushing forces limited?
104. Is the body weight used during pulling and pushing?
105. Are trolleys fitted with handgrips?
106. Do trolleys have two swivel wheels?
107. Are the floors hardened and even?

ENVIRONMENTAL FACTORS (CHAPTER 4)

NOISE

Guidelines on Noise

108. Is the noise level below 80 decibels?
109. Is annoyance due to noise avoided?
110. Are rooms perhaps too quiet?

Noise Reduction at Source

111. Has a low-noise working method been chosen?
112. Are quiet machines used?
113. Are machines regularly maintained?

114. Are noisy machines enclosed?

Noise Reduction through Workplace Design and Work Organization

115. Is noisy work separated from quiet work?
116. Is there an adequate distance from the source of noise?
117. Is the ceiling used for noise absorption?
118. Are acoustic screens used?

Hearing Conservation

119. Are hearing conservation measures suited both to the noise and to the user?

VIBRATION

Guidelines for Vibration

120. Are health and safety risks from vibration avoided?
121. Are shocks and jolts prevented?

Preventing Vibration

122. Is vibration tackled at source?
123. Are machines regularly maintained?
124. Is the transmission of vibration prevented?
125. Are measures at the individual level applied only as a last resort?

ILLUMINATION

Guidelines on Light Intensity

126. Is the light intensity for orientation tasks in the range of 20–200 lux?
127. Is the light intensity for normal activities in the range of 250–750 lux?
128. Is the light intensity for special applications in the range of 750–5000 lux?

Guidelines on Brightness Differences

129. Are large brightness differences in the visual field avoided?
130. Are the brightness differences between task area, close surroundings and wider surroundings limited?

Guidelines for the Color of the Light

131. Are too cold and too warm colors for indoor lighting avoided?

Improved Lighting

132. Is the information easily legible?
133. Is ambient lighting combined with localized lighting?

134. Is daylight also used for ambient lighting?
135. Are direct light sources properly screened?
136. Are reflections and shadows avoided?
137. Is diffuse direct light used?
138. Is flicker from fluorescent tubes avoided?

CLIMATE

Guidelines on Thermal Comfort

139. Are people able to control the climate themselves?
140. Is the air temperature suited to the physical demands of the task?
141. Is the air prevented from becoming either too dry or too humid?
142. Are hot radiating surfaces avoided?
143. Are drafts prevented?

Guidelines on Heat and Cold

144. Is the exposure to extreme hot or cold environments avoided?
145. Are materials that have to be touched neither too cold nor too hot?

Climate Control

146. Are equally heavy tasks grouped in the same room?
147. Are the physical demands of the task adjusted to the external climate?
148. Is the air velocity optimal?
149. Is undesirable radiation prevented?
150. Is the time spent in hot or cold environments limited?
151. Is special clothing used when spending long periods in hot or cold environments?

CHEMICAL SUBSTANCES

Guidelines on Chemical Substances

152. Is the concentration of chemical substances in air subject to limits (TLVs or other)?
153. Are carcinogenic substances avoided?
154. Are peak exposures prevented?
155. Is exposure to mixtures of substances prevented?
156. Is the concentration of chemical substances as far below the TLVs as possible?
157. Are packages of chemicals appropriately labeled?

Measures at Source

158. Can the source be removed?
159. Can releases from the source be reduced?
160. Can the source be isolated?

Ventilation

161. Are chemical substances extracted directly at the source?
162. Is the air exhaust system efficient?
163. Has attention been paid to climate at workplaces where exhaust and ventilation are used?
164. Are sufficient air changes provided?

Measures at the Individual Level

165. Are organizational measures possible?
166. Is personal protective equipment available?
167. Are dust masks used only as protection against coarse dust?
168. Are protective clothing and gloves available?
169. Is attention paid to personal hygiene?

FACTORS RELATED TO INFORMATION AND OPERATION (CHAPTER 3)

THE USER

170. Is the user population defined as accurately as possible?
171. Are cultural differences taken into account?

INFORMATION

Visual Information

172. Have texts with only capitals been avoided?
173. Have familiar typefaces been chosen?
174. Has confusion between characters been avoided?
175. Has the correct character size been chosen?
176. Are longer lines more widely spaced?
177. Is the contrast good?
178. Are the diagrams easily understood?
179. Have pictograms been properly used?

Hearing

180. Are sounds reserved for warning signals?
181. Has the correct pitch been chosen?
182. Is synthesized speech adjustable?

Other Senses

183. Are the uses of taste, smell and temperature restricted to warning signals?
184. Is the sense of touch used for feedback from controls?
185. Are different senses used for simultaneous information?

CONTROLS FOR OPERATION

FIXED CONTROLS

186. Is a familiar keyboard layout used?
187. Is the number of function keys limited?
188. Is the type of cursor control suited to the task?
189. Is the mouse not used exclusively?
190. Are touch screens used to facilitate operation by inexperienced users?
191. Are pedals used only where the use of the hands is inconvenient?
192. Is unintentional operation avoided?
193. Are labels or symbols properly used?
194. Is the use of color limited?
195. Is the direction of movement consistent with expectation?
196. Is the objective clear from the position of the controls?

Wireless, Remote and Hands-Free Controls

197. Are remote controls used to give the user more freedom?
198. Are hands-free controls avoided for precise data?
199. Is speech recognition limited to quiet environments?

DIALOGUES

User Friendliness

200. Is the dialogue suitable to the user's task and skill level?
201. Does the dialogue make clear what the user should do next?
202. Will the user be in control of the pace and sequence dialogue?
203. Is the dialogue consistent?
204. Is the dialogue forgiving?
205. Is the dialogue suitable for individualization?
206. Is the dialogue suitable for learning?

Different Forms of Dialogue

207. Have menus been used for users with little knowledge or experience?
208. Are the limitations of an input form known?
209. Has command language been restricted to experienced users?
210. Is direct manipulation consistent?
211. Have the disadvantages of natural language been recognized?

Help

212. Is the type of help suited to the level of the user?

WEBSITE DESIGN

213. Can the behavior of the user on the website be predicted?

214. Is the presentation of the website independent of the user's hardware and browser?
215. Is the content understandable and navigable?

MOBILE INTERACTION

216. Is mobile interaction used for simple content?
217. Are screen layout and navigation adapted to the needs of the user?

VIRTUAL REALITY

218. Is an appropriate way of manipulating objects chosen?
219. Is the playability of a game improved by enlarging the user experience?

CONCLUDING FACTOR

220. Would you wish to carry out the task yourself?

7 Sources of Additional Information

This chapter provides information to readers wishing to learn more about ergonomics. We first cite some international general and more specialized ergonomics books. Next, the most relevant scientific ergonomics journals are presented. Then some useful websites are mentioned. This chapter concludes with an extensive list of international ergonomics ISO standards.

7.1 INTERNATIONAL BOOKS ON ERGONOMICS

The background to the guidelines and advice given in this book can be found in the literature listed below. The more general sources appear first, followed by the more specialist sources grouped according to the chapters of this book.

7.1.1 GENERAL ERGONOMICS BOOKS

Bridger, R.S., 2003, *Introduction to Ergonomics,* McGraw-Hill, ISBN 0415273781

Helander, M., 2005, A *Guide to Human Factors and Ergonomics, Second Edition*, CRC Press, ISBN 0415282489

Marras, W.S. and Karwowski W., 2006, Fundamentals and Assessment Tools for Occupational Ergonomics. *The Occupational Ergonomics Handbook, Second Edition*, Taylor & Francis, ISBN 0849319374

Marras, W.S. and Karwowski W., 2006, Interventions, Controls, and Applications in Occupational Ergonomics. *The Occupational Ergonomics Handbook, Second Edition*, Taylor & Francis, ISBN 0849319382

Karwowski W. (ed.), 2005, *Handbook of Standards and Guidelines in Ergonomics and Human Factors*, Lawrence Erlbaum Associates, ISBN 0805841296

Karwowski, W. (ed.), 2006, *International Encyclopedia of Ergonomics and Human Factors,* Taylor & Francis, ISBN 041530430X

Konz, S. and Johnson, S., 2004, *Work Design: Industrial Ergonomics, Sixth Edition,* Holcomb Hathaway, ISBN 9781890871482

Kroemer, K.H.E. and Grandjean, E. (eds.), 1997, *Fitting the Task to the Human: A Textbook of Occupational Ergonomics, Fifth Edition*, Taylor & Francis, ISBN 0748406654

Kroemer, K.H.E., 2005, *"Extra-Ordinary" Ergonomics: How to Accommodate Small and Big Persons, The Disabled and Elderly, Expectant Mothers, and Children*, Taylor & Francis, ISBN 0849336686

Norman, D.A., 2002, *The Design of Everyday Things*, Basic Books, ISBN 9780465067107

Salvendy, G. (ed.), 2006, *Handbook of Human Factors and Ergonomics, Third Edition*, Wiley ISBN 9780471449171

The Eastman Kodak Company, 2003, *Kodak's Ergonomic Design for People at Work, Second Edtion*, Wiley, ISBN 9780471418634

Wickens, C.D., Gordon, S.E., and Liu, Y., 2003, *An Introduction to Human Factors Engineering, Second Edition*, Prentice Hall, ISBN 9780131837362

7.1.2 BOOKS ON POSTURE AND MOVEMENT

Cacha, C.A., 1999, *Ergonomics and Safety in Hand Tool Design,* CRC Press, ISBN 9781566703086

Chaffin, D.B., Andersson, G.B.J., and Martin, B.J., 2006, *Occupational Biomechanics, Fourth Edition,* John Wiley, ISBN 9780471723431

Delleman, N.J., Haslegrave, C.M., and Chaffin, D.B., 2004, *Working Postures and Movements,* CRC Press, ISBN 9780415279086

Pheasant, S. and Haslegrave, C., 2006, *Bodyspace: Anthropometry, Ergonomics and the Design of Work, Third Edition,* CRC Press, ISBN 0415285208

Kumar, S. (ed.), 1999, *Biomechanics in Ergonomics,* CRC Press, ISBN 0748407049

Nordin, M., Andersson, G.B.J., and Pope, M.H., 2006, *Musculoskeletal Disorders in the Workplace: Principles and Practice,* Mosby, ISBN 9780323026222

Putz-Anderson, V. (ed.), 2006, *Cumulative Trauma Disorders, Second Edition,* CRC Press, ISBN 9780415232234

Tilley, A.R. and Henry Dreyfuss Associates, 2001, *The Measure of Man and Woman: Human Factors in Design,* Wiley, ISBN 9780471099550

7.1.3 BOOKS ON INFORMATION AND OPERATION

Chen, F., 2006, *Designing Human Interface in Speech Technology,* Springer, ISBN 0387241558

Dix, A., Finlay, J.E., Abowd, G.D., and Beale, R., 2004, *Human-Computer Interaction, Third Edition,* Prentice Hall, ISBN 0130461091

Jones, M. and Marsden, G., 2006, *Mobile Interaction Design,* John Wiley & Sons, ISBN 0470090898

Love, S., 2005, *Understanding Mobile Human-Computer Interaction,* Butterworth-Heinemann, ISBN 0750663529

Moggridge, B., 2006, *Designing Interactions,* The MIT Press, ISBN 0262134748

Nielsen, J. and Loranger, H., 2006, *Prioritizing Web Usability,* New Riders Press, ISBN 0321350316

Shneiderman, B. and Plaisant, C., 2004, *Designing the User Interface: Strategies for Effective Human-Computer Interaction Fourth Edition,* Addison Wesley, ISBN 0321197860

Sherman, W.R. and Craig, A.B., 2003, *Understanding Virtual Reality,* Elsevier, ISBN 1558603530.

7.1.4 BOOKS ON ENVIRONMENTAL FACTORS

Anshel, J., 2005, *Visual Ergonomics Handbook,* CRC Press, ISBN 1566706823

Boyce, P.R., 2003, *Human Factors in Lighting, Second Edition,* Taylor & Francis, ISBN 0748409505

Mansfield, N.J., 2004, *Human Response to Vibration,* CRC Press, ISBN 041528239X

South, T., 2004, *Managing Noise and Vibration at Work: A Practical Guide to Assessment, Measurement and Control,* Butterworth-Heinemann, ISBN 0750663421

Gardiner, K. and Harrington, J.M. (eds.), 2005, *Occupational Hygiene, Third Edition,* Blackwell Publishing Professional, ISBN 1405106212

Parsons, K.C., 2002, *Human Thermal Environments. The Effects of Hot, Moderate, and Cold Environments on Human Health, Comfort and Performance, Second Edition,* Taylor & Francis, ISBN 0415237939

7.1.5 BOOKS ON WORK ORGANIZATION, JOBS AND TASKS

Hackman, J.R. and Oldham, G.R., 1980, *Work Redesign,* Addison Wesley) ISBN 0201027798

Hendrick, H. and Kleiner, B. (eds.), 2002, *Macroergonomics: Theory, Methods, and Applications,* Lawrence Erlbaum Associates, ISBN 0805831916

Parker, S. and Wall, T., 1998, *Job and Work Design: Organizing Work* to *Promote Well-Being and Effectiveness,* SAGE Publications, ISBN 0761904204

Tosi, H.L. and Mero, N.P., 2003, *The Fundamentals of Organizational Behavior: What Managers Need to Know,* Blackwell Publishing Professional, ISBN 1405100745

Anderson, N., Ones, D.S, Sinangil, H.K., and Viswesvaran, C., 2002, *Handbook of Industrial, Work and Organizational Psychology, Volume 2: Organizational Psychology,* Sage Publications, ISBN 0761964894

7.1.6 BOOKS ON THE ERGONOMIC APPROACH

Noro, K. and Imada, A. (eds.), 1991, *Participatory Ergonomics,* Taylor & Francis, ISBN 0850663282

Oxenburgh, M.S., Marlow, P.S.P., and Oxenburgh, A., 2004, *Increasing Productivity and Profit through Health and Safety: The Financial Returns from a Safe Working Environment,* CRC Press, ISBN 0415243319

Stanton, N.A., Hedge, A., Brookhuis, K., and Hendrick, H. (eds.), 2004, *Handbook of Human Factors and Ergonomics Methods,* CRC Press, ISBN 0415287006

Stanton, N.A., Salmon, P.M., Walker, G.H., Baber, C., and Jenkins, D.P., 2005, *Human Factors Methods: A Practical Guide for Engineering and Design,* Ashgate Publishing, ISBN 0754646610

Wilson, J.R. and Corlett, N. (2005), *Evaluation of Human Work, Third Edition,* CRC Press, ISBN 0415267579

7.2 SCIENTIFIC JOURNALS ON ERGONOMICS

Applied Ergonomics (Amsterdam: Elsevier), ISSN 0003-6870

Ergonomia. An International Journal of Ergonomics and Human Factors (Warsaw: Polish Academy of Sciences), ISSN 0137-4990

*Ergonomia (Italy). (*Milan: Moretti Vitali Editori), ISSN 0014-0120

Ergonomics (London: Taylor & Francis), ISSN 0014-0139

Ergonomics Abstracts (London: Taylor & Francis), ISSN 1464-5084

Human Factors (Santa Monica, California: Human Factors and Ergonomics Society), ISSN 0018-7208

Human Factors and Ergonomics in Manufacturing (New York: John Wiley), ISSN 1090-8471

International Journal of Cognitive Ergonomics (Mahwah, New Jersey: Lawrence Erlbaum Associates), ISSN 1088-6362 (until 2000)

International Journal of Human Factors Modelling and Simulation (Geneva: Interscience Publishers), ISSN 1742-5549

International Journal of Industrial Ergonomics (Amsterdam: Elsevier), ISSN 0169-8141

International Journal of Occupational Safety and Ergonomics (Warsaw: Central Institute for Labour Protection), ISSN 1080-3548

Japanese Journal of Ergonomics (Tokyo: Business Center for Academic Societies Japan), ISSN 0549-4974

Occupational Ergonomics (Amsterdam: IOS), ISSN 1359-9364

Theoretical Issues in Ergonomics Science: TIES (London: Taylor & Francis), ISSN 1463-922X

Travail Humain (Paris: Presses Universitaires de France), ISSN 0041-1868

Zeitschrift für Arbeitswissenschaft (Stuttgart: Ergonomia Verlag), ISSN 0340-2444

Zentralblatt für Arbeitsmedizin, Arbeitsschutz und Ergonomie (Heidelberg: Curt Haefner Verlag), ISSN 0944-2502

7.3   USEFUL WEBSITES

The websites listed below are just a few of the available sites devoted to ergonomics. However, the links found in these sites will allow you to gain access to most of the other available sites.

Bad ergonomics designs

"Baddesigns" is a website that shows examples of bad ergonomics in everyday life, including suggestions for design improvements.

http://www.baddesigns.com

Board of Certification in Professional Ergonomics (BCPE)

The BCPE is the certifying body in the USA for individuals whose education and experience indicate broad expertise in the practice of human factors/ergonomics. It awards the credential Certified Professional Ergonomist (CPE).

http://www.bcpe.org/

Centre for Registration of European Ergonomists (CREE)

The CREE is a cooperation of some 15 European ergonomics societies that specifies the standards of knowledge and practical experience that define the European Ergonomist.

http://www.eurerg.org

Ergonomics Abstracts database

Ergonomics Abstracts is a database that includes bibliographical information and abstracts of ergonomics books and ergonomics articles in journals and conference proceedings.

http://www.tandf.co.uk/ergo-abs/

Ergonomics for schools

This website promotes learning about ergonomics among secondary school students and their teachers.

www.ergonomics4schools.com

Ergonomics Society, UK

The Ergonomics Society of the United Kingdom was established in 1949 and is the oldest ergonomics society worldwide.

http://www.ergonomics.org.uk

Ergoweb
> Ergoweb is the website of the company Ergoweb Inc. with ergonomics news. It focuses on physical ergonomics and the U.S.
>> http://www.ergoweb.com

Federation of European Ergonomics Societies (FEES)
> The FEES is a regional network of the International Ergonomics Association that comprises some 15 European ergonomics societies representing about 4,000 members.
>> http://www.fees-network.org

Human Factors and Ergonomics Society, USA
> The Human Factors and Ergonomics Society is one of the largest ergonomics societies worldwide with around 4,500 members.
>> http://www.hfes.org

International Ergonomics Association (IEA)
> The IEA comprises approximately 40 member societies representing about 19,000 ergonomists worldwide.
>> http://www.iea.cc

Usernomics
> Usernomics is a company focusing on the usability of software, hardware, and the workplace. The website include useful Internet links.
>> http://www.usernomics.com

7.4   INTERNATIONAL ISO STANDARDS ON ERGONOMICS

Below is a list of published standards of the International Organization for Standardization in the field of ergonomics. The standards were prepared by the technical committees ISO TC 159 "Ergonomics" (all published standards), ISO TC 199 "Safety of machinery" (all published standards), ISO TC 108 Subcommittee 4 "Human exposure to mechanical vibration and shock "(all published standards), ISO JTC1/SC35 "User interfaces" (all published standards), ISO TC 43/SC1 "Noise" (selection of published standards), and CIE "International Commission on Illumination" (selection of published standards). These standards are the result of a standardization process in which interested parties, usually including ergonomics experts, seek consensus.

In the lists below, the standards are first classified according to the topics of the chapters of this book. A standard can be listed under more than one topic. Then the standards are listed (with year published) according to ascending identification number.

Many ergonomics standards are still under development within the above technical committees. Furthermore, in other technical committees, standards have been developed and are under development in which ergonomics is applied to specific work environments like IT systems, aircraft, road vehicles, tractors, forest machinery, petroleum and natural gas industry, textile industry, building industry, earth moving machinery, etc.

An up-to-date list of ISO standards can be found on www. iso.org.

Abbreviations mentioned in this section have the following meanings: Amd. = Amendment; CIE= International Commission on Illumination; Cor. = Correction; IEC = International Electrotechnical Commission; JTC = Joint Technical Committee; PAS = Publicly Available Specification, SC = Subcommittee, TC = Technical Committee, TR = Technical Report; TS = Technical Specification.

7.4.1 GENERAL ERGONOMICS STANDARDS

General design:
    ISO 6385

Workplace, equipment, and product design:
    ISO 9241-5, ISO 11064-1, ISO 11064-2, ISO 11064-3, ISO 11064-4, ISO 20282-1, ISO 20282-2

Safety of machinery:
    ISO 12100-1, ISO 12100-2, ISO 13849-1, ISO 13849-2, ISO/TR 13849-100, ISO 13850, ISO 13851, ISO 13852, ISO 13853, ISO 13854, ISO 13855, ISO 13856-1, ISO 13856-2, ISO 13856-3, ISO 14118, ISO 14119, ISO 14120, ISO 14121, ISO 14122-1, ISO 14122-2, ISO 14122-3, ISO 14122-4, ISO 14123-1, ISO 14123-2, ISO 14159, ISO/TR 18569, ISO 214

7.4.2 STANDARDS ON POSTURE AND MOVEMENT

Biomechanics:
    ISO 1503, ISO 11226, ISO 11228-1, ISO 20646

Anthropometrics:
    ISO 7250, ISO 14738, ISO 15534-1, ISO 15534-2, ISO 15534-3, ISO 15535, ISO 15536-1, ISO 15537, ISO 20685

7.4.3 STANDARDS ON INFORMATION AND OPERATION

General:
    ISO 9241-1, ISO 9241-2, ISO 9241-3, ISO 9241-4, ISO 9241-5, ISO 9241-6, ISO 9241-7, ISO 9241-8, ISO 9241-9, ISO 11428, ISO 13406-1, ISO 13406-2, ISO 16071, ISO 16982, ISO 18152, ISO 18529

User interfaces:
    ISO/IEC 9995-1, ISO/IEC 9995-2, ISO/IEC 9995-3, ISO/IEC 9995-4, ISO/IEC 9995-5, ISO/IEC 9995-6, ISO/IEC 9995-7, ISO/IEC 9995-8, ISO/IEC 10741-1, ISO/IEC 10741-1:1995/Amd 1, ISO/IEC 11581-1, ISO/IEC 11581-2,

ISO/IEC 11581-3, ISO/IEC 11581-4, ISO/IEC 11581-5, ISO/IEC 11581-6, ISO/IEC 13251, ISO/IEC 14754, ISO/IEC 14755, ISO/IEC 15411, ISO/IEC 15412, ISO/IEC TR 15440, ISO/IEC 15897

Software:

ISO 9241-11 ISO 9241-12 ISO 9241-13 ISO 9241-14 ISO 9241-15 ISO 9241-16 ISO 9241-17, ISO 9241-110, ISO 13407, ISO 14915-1, ISO 14915-2, ISO 14915-3

Displays and controls

ISO 7731, ISO 9241-4, ISO 9355-1, ISO 9355-2, ISO 9921, ISO 11428, ISO 11429, ISO 19358

7.4.4 STANDARDS ON ENVIRONMENTAL FACTORS

General:

ISO 9241-6, ISO 11064-6

Noise:

ISO 1996-1, ISO 1996-2, ISO 1996-3, ISO 1999, ISO 4869-1, ISO 4869-2, ISO/TR 4869-3, ISO/TR 4869-4, ISO/TR 4869-5, ISO 7731, ISO 11429, ISO 15664, ISO/TS 15666, ISO 15667, ISO 17624

Vibration:

ISO 2631-1, ISO 2631-2, ISO 2631-4, ISO 2631-5, ISO 5349-1, ISO 5349-2, ISO 5805, ISO 5982, ISO 6897, ISO 8727, ISO, 9996, ISO 10068, ISO 10227, ISO 10326-1, ISO 10326-2, ISO 10819, ISO 13090-1, ISO 13091-1, ISO 13092-2, ISO 13753, ISO 14835-1, ISO 14835-2, ISO/TS 15694

Illumination:

ISO/CIE 8995-1, ISO/CIE 8995-3, ISO 9241-7

Climate:

ISO 7243, ISO 7726, ISO 7730, ISO 7933, ISO 8996, ISO 9886, ISO 9920, ISO 10551, ISO/TR 11079, ISO 11399, ISO 12894, ISO 13731, ISO 13732-1, ISO/TS 13732-2, ISO 14415, ISO 14505-2, ISO 14505-3, ISO 15265

7.4.5 STANDARDS ON WORK ORGANIZATION, JOBS AND TASKS

General:

ISO 6385, ISO 9241-2

Mental workload:

ISO 10075, ISO 10075-2, ISO 10075-3

7.4.6 STANDARDS ON THE ERGONOMICS APPROACH

ISO 6385, ISO 9241-1, ISO 11064-1, ISO 11064-7, ISO 13407

7.4.7 LIST OF STANDARDS

ISO 1503:1977 Geometrical orientation and directions of movements

ISO 1996-1:2003 Acoustics — Description, measurement and assessment of environmental noise — Part 1: Basic quantities and assessment procedures

ISO 1996-2:1987/Amd 1:1998 Acoustics — Description and measurement of environmental noise — Part 2: Acquisition of data pertinent to land use

ISO 1996-3:1987 Acoustics — Description and measurement of environmental noise — Part 3: Application to noise limits

ISO 1999:1990 Acoustics — Determination of occupational noise exposure and estimation of noise-induced hearing impairment

ISO 2631-1:1997 Mechanical vibration and shock — Evaluation of human exposure to whole-body vibration — Part 1: General requirements

ISO 2631-2:2003 Mechanical vibration and shock — Evaluation of human exposure to whole-body vibration — Part 2: Vibration in buildings (1 Hz to 80 Hz)

ISO 2631-4:2001 Mechanical vibration and shock — Evaluation of human exposure to whole-body vibration — Part 4: Guidelines for the evaluation of the effects of vibration and rotational motion on passenger and crew comfort in fixed-guideway transport systems

ISO 2631-5:2004 Mechanical vibration and shock — Evaluation of human exposure to whole-body vibration — Part 5: Method for evaluation of vibration containing multiple shocks

ISO 4869-1:1990 Acoustics — Hearing protectors — Part 1: Subjective method for the measurement of sound attenuation

ISO 4869-2:1994/Cor 1:2006 Acoustics — Hearing protectors — Part 2: Estimation of effective A-weighted sound pressure levels when hearing protectors are worn

ISO/TR 4869-3:1989 Acoustics — Hearing protectors — Part 3: Simplified method for the measurement of insertion loss of ear-muff type protectors for quality inspection purposes

ISO/TR 4869-4:1998 Acoustics — Hearing protectors — Part 4: Measurement of effective sound pressure levels for level-dependent sound-restoration ear-muffs

ISO/TS 4869-5:2006 Acoustics — Hearing protectors — Part 5: Method for estimation of noise reduction using fitting by inexperienced test subjects

ISO 5349-1:2001 Mechanical vibration — Measurement and evaluation of human exposure to hand-transmitted vibration — Part 1: General requirements

ISO 5349-2:2001 Mechanical vibration — Measurement and evaluation of human exposure to hand-transmitted vibration — Part 2: Practical guidance for measurement at the workplace

ISO 5805:1997 Mechanical vibration and shock — Human exposure — Vocabulary

ISO 5982:2001 Mechanical vibration and shock — Range of idealized values to characterize seated-body biodynamic response under vertical vibration

ISO 6385:2004 Ergonomic principles in the design of work systems

ISO 6897:1984 Guidelines for the evaluation of the response of occupants of fixed structures, especially buildings and off-shore structures, to low-frequency horizontal motion (0,063 to 1 Hz)

ISO 7243:1989 Hot environments — Estimation of the heat stress on working man, based on the WBGT-index (wet bulb globe temperature)

ISO 7250:1996 Basic human body measurements for technological design

ISO 7726:1998 Ergonomics of the thermal environment — Instruments for measuring physical quantities

ISO 7730:2005 Ergonomics of the thermal environment — Analytical determination and interpretation of thermal comfort using calculation of the PMV and PPD indices and local thermal comfort criteria

ISO 7731:2003 Danger signals for work places — Auditory danger signals

ISO 7933:2004 Ergonomics of the thermal environment — Analytical determination and interpretation of heat stress using calculation of the predicted heat strain

ISO 8727:1997 Mechanical vibration and shock — Human exposure — Biodynamic coordinate systems

ISO/CIE 8995-1:2002/Cor 1:2005 Lighting of work places — Part 1: Indoor

ISO/CIE 8995-3:2006 Lighting of work places — Part 3: Lighting requirements for safety and security of outdoor work places

ISO 8996:2004 Ergonomics of the thermal environment — Determination of metabolic rate

ISO 9241-1:1997/Amd 1:2001 Ergonomic requirements for office work with visual display terminals (VDTs) — Part 1: General introduction

ISO 9241-2:1992 Ergonomic requirements for office work with visual display terminals (VDTs) — Part 2: Guidance on task requirements

ISO 9241-3:1992/Amd 1:2000 Ergonomic requirements for office work with visual display terminals (VDTs) — Part 3: Visual display requirements

ISO 9241-4:1998/Cor 1:2000 Ergonomic requirements for office work with visual display terminals (VDTs) — Part 4: Keyboard requirements

ISO 9241-5:1998 Ergonomic requirements for office work with visual display terminals (VDTs) — Part 5: Workstation layout and postural requirements

ISO 9241-6:1999 Ergonomic requirements for office work with visual display terminals (VDTs) — Part 6: Guidance on the work environment

ISO 9241-7:1998 Ergonomic requirements for office work with visual display terminals (VDTs) — Part 7: Requirements for display with reflections

ISO 9241-8:1997 Ergonomic requirements for office work with visual display terminals (VDTs) — Part 8: Requirements for displayed colours

ISO 9241-9:2000 Ergonomic requirements for office work with visual display terminals (VDTs) — Part 9: Requirements for non-keyboard input devices

ISO 9241-11:1998 Ergonomic requirements for office work with visual display terminals (VDTs) — Part 11: Guidance on usability

ISO 9241-12:1998 Ergonomic requirements for office work with visual display terminals (VDTs) — Part 12: Presentation of information

ISO 9241-13:1998 Ergonomic requirements for office work with visual display terminals (VDTs) — Part 13: User guidance

ISO 9241-14:1997 Ergonomic requirements for office work with visual display terminals (VDTs) — Part 14: Menu dialogues

ISO 9241-15:1997 Ergonomic requirements for office work with visual display terminals (VDTs) — Part 15: Command dialogues

ISO 9241-16:1999 Ergonomic requirements for office work with visual display terminals (VDTs) — Part 16: Direct manipulation dialogues

ISO 9241-17:1998 Ergonomic requirements for office work with visual display terminals (VDTs) — Part 17: Form filling dialogues

ISO 9241-110:2006 Ergonomics of human-system interaction — Part 110: Dialogue principles

ISO 9355-1:1999 Ergonomic requirements for the design of displays and control actuators — Part 1: Human interactions with displays and control actuators

ISO 9355-2:1999 Ergonomic requirements for the design of displays and control actuators — Part 2: Displays

ISO 9886:2004 Ergonomics: Evaluation of thermal strain by physiological measurements

ISO 9920:1995 Ergonomics of the thermal environment — Estimation of the thermal insulation and evaporative resistance of a clothing ensemble

ISO 9921:2003 Ergonomic assessment of speech communication — Part 1: Speech interference level and communication distances for persons with normal hearing capacity in direct communication (SIL method)

ISO/IEC 9995-1:2006 Information technology — Keyboard layouts for text and office systems — Part 1: General principles governing keyboard layouts

ISO/IEC 9995-2:2002 Information technology — Keyboard layouts for text and office systems — Part 2: Alphanumeric section

ISO/IEC 9995-3:2002 Information technology — Keyboard layouts for text and office systems — Part 3: Complementary layouts of the alphanumeric zone of the alphanumeric section

ISO/IEC 9995-4:2002 Information technology — Keyboard layouts for text and office systems — Part 4: Numeric section

ISO/IEC 9995-5:2006 Information technology — Keyboard layouts for text and office systems — Part 5: Editing section

ISO/IEC 9995-6:2006 Information technology — Keyboard layouts for text and office systems — Part 6: Function section

ISO/IEC 9995-7:2002 Information technology — Keyboard layouts for text and office systems — Part 7: Symbols used to represent functions

ISO/IEC 9995-8:2006 Information technology — Keyboard layouts for text and office systems — Part 8: Allocation of letters to the keys of a numeric keypad

ISO 9996:1996 Mechanical vibration and shock — Disturbance to human activity and performance — Classification

ISO 10068:1998 Mechanical vibration and shock — Free, mechanical impedance of the human hand-arm system at the driving point

ISO 10075:1991 Ergonomic principles related to mental work-load — General terms and definitions

ISO 10075-2:1996 Ergonomic principles related to mental workload — Part 2: Design principles

ISO 10075-3:2004 Ergonomic principles related to mental workload — Part 3: Principles and requirements concerning methods for measuring and assessing mental workload

ISO 10227:1996 Human/human surrogate impact (single shock) testing and evaluation — Guidance on technical aspects

ISO 10326-1:1992 Mechanical vibration — Laboratory method for evaluating vehicle seat vibration — Part 1: Basic requirements

ISO 10326-2:2001 Mechanical vibration — Laboratory method for evaluating vehicle seat vibration — Part 2: Application to railway vehicles

ISO 10551:1995 Ergonomics of the thermal environment — Assessment of the influence of the thermal environment using subjective judgement scales

ISO/IEC 10741-1:1995 Information technology — User system interfaces — Dialogue interaction — Part 1: Cursor control for text editing

ISO/IEC 10741-1:1995/Amd 1:1996 Macro cursor control

ISO 10819:1996 Mechanical vibration and shock — Hand-arm vibration — Method for the measurement and evaluation of the vibration transmissibility of gloves at the palm of the hand

ISO 11064-1:2000 Ergonomic design of control centres — Part 1: Principles for the design of control centres

ISO 11064-2:2000 Ergonomic design of control centres — Part 2: Principles for the arrangement of control suites

ISO 11064-3:1999/Cor 1:2002 Ergonomic design of control centres — Part 3: Control room layout

ISO 11064-4:2004 Ergonomic design of control centres — Part 4: Layout and dimensions of workstations

ISO 11064-6:2005 Ergonomic design of control centres — Part 6: Environmental requirements for control centres

ISO 11064-7:2006 Ergonomic design of control centres — Part 7: Principles for the evaluation of control centres

ISO/TR 11079:1993 Evaluation of cold environments — Determination of requisite clothing insulation (IREC)

ISO 11226:2000 Ergonomics — Evaluation of static working postures

ISO 11228-1:2003 Ergonomics — Manual Handling — Part 1: Lifting and carrying

ISO 11399:1995 Ergonomics of the thermal environment — Principles and application of relevant International Standards

ISO 11428:1996 Ergonomics — Visual danger signals — General requirements, design and testing

ISO 11429:1996 Ergonomics — System of auditory and visual danger and information signals

ISO/IEC 11581-1:2000 Information technology — User system interfaces and symbols — Icon symbols and functions — Part 1: Icons — General

ISO/IEC 11581-2:2000 Information technology — User system interfaces and symbols — Icon symbols and functions — Part 2: Object icons

ISO/IEC 11581-3:2000 Information technology — User system interfaces and symbols — Icon symbols and functions — Part 3: Pointer icons

ISO/IEC 11581-5:2004 Information technology — User system interfaces and symbols — Icon symbols and functions — Part 5: Tool icons

ISO/IEC 11581-6:1999 Information technology — User system interfaces and symbols — Icon symbols and functions — Part 6: Action icons

ISO 12100-1:2003 Safety of machinery — Basic concepts, general principles for design — Part 1: Basic terminology, methodology

ISO 12100-2:2003 Safety of machinery — Basic concepts, general principles for design — Part 2: Technical principles

ISO 12894:2001 Ergonomics of the thermal environment — Medical supervision of individuals exposed to extreme hot or cold environments

ISO 13090-1:1998 Mechanical vibration and shock — Guidance on safety aspects of tests and experiments with people — Part 1: Exposure to whole-body mechanical vibration and repeated shock

ISO 13091-1:2001 Mechanical vibration — Vibrotactile perception thresholds for the assessment of nerve dysfunction — Part 1: Methods of measurement at the fingertips

ISO 13091-2:2003 Mechanical vibration — Vibrotactile perception thresholds for the assessment of nerve dysfunction — Part 2: Analysis and interpretation of measurements at the fingertips

ISO/IEC 13251:2004 Collection of graphical symbols for office equipment

ISO 13406-1:1999 Ergonomic requirements for work with visual displays based on flat panels — Part 1: Introduction

ISO 13406-2:2001 Ergonomic requirements for work with visual displays based on flat panels — Part 2: Ergonomic requirements for flat panel displays

ISO 13407:1999 Human-centered design processes for interactive systems

ISO 13731:2001 Ergonomics of the thermal environment — Vocabulary and symbols

ISO 13732-1:2006 Ergonomics of the thermal environment — Methods for the assessment of human responses to contact with surfaces — Part 1: Hot surfaces

ISO/TS 13732-2:2001 Ergonomics of the thermal environment — Methods for the assessment of human responses to contact with surfaces — Part 2: Human contact with surfaces at moderate temperature

ISO 13732-3:2005 Ergonomics of the thermal environment — Methods for the assessment of human responses to contact with surfaces — Part 3: Cold surfaces

ISO 13753:1998 Mechanical vibration and shock — Hand-arm vibration — Method for measuring the vibration transmissibility of resilient materials when loaded by the hand-arm system

ISO 13849-1:2006 Safety of machinery — Safety-related parts of control systems — Part 1: General principles for design

ISO 13849-2:2003 Safety of machinery — Safety-related parts of control systems — Part 2: Validation

ISO/TR 13849-100:2000 Safety of machinery — Safety-related parts of control systems — Part 100: Guidelines for the use and application of ISO 13849-1

ISO 13850:2006 Safety of machinery — Emergency stop — Principles for design

ISO 13851:2002 Safety of machinery — Two-hand control devices — Functional aspects and design principles

ISO 13852:1996 Safety of machinery — Safety distances to prevent danger zones being reached by the upper limbs

ISO 13853:1998 Safety of machinery — Safety distances to prevent danger zones being reached by the lower limbs

ISO 13854:1996 Safety of machinery — Minimum gaps to avoid crushing of parts of the human body

ISO 13855:2002 Safety of machinery — Positioning of protective equipment with respect to the approach speeds of parts of the human body

ISO 13856-1:2001 Safety of machinery — Pressure-sensitive protective devices — Part 1: General principles for design and testing of pressure-sensitive mats and pressure-sensitive floors

ISO 13856-2:2005 Safety of machinery — Pressure-sensitive protective devices — Part 2: General principles for the design and testing of pressure-sensitive edges and pressure-sensitive bars

ISO 13856-3:2006 Safety of machinery — Pressure-sensitive protective devices — Part 3: General principles for the design and testing of pressure-sensitive bumpers, plates, wires and similar devices

ISO 14118:2000 Safety of machinery — Prevention of unexpected start-up

ISO 14119:1998 Safety of machinery — Interlocking devices associated with guards — Principles for design and election

ISO 14120:2002 Safety of machinery — Guards — General requirements for the design and construction of fixed and movable guards

ISO 14121:1999 Safety of machinery — Principles of risk assessment

ISO 14122-1:2001 Safety of machinery — Permanent means of access to machinery — Part 1: Choice of fixed means of access between two levels

ISO 14122-2:2001 Safety of machinery — Permanent means of access to machinery — Part 2: Working platforms and walkways

ISO 14122-3:2001 Safety of machinery — Permanent means of access to machinery — Part 3: Stairs, stepladders and guard-rails

ISO 14122-4:2004 Safety of machinery — Permanent means of access to machinery — Part 4: Fixed ladders

ISO 14123-1:1998 Safety of machinery — Reduction of risks to health from hazardous substances emitted by machinery — Part 1: Principles and specifications for machinery manufacturers

ISO 14123-2:1998 Safety of machinery — Reduction of risks to health from hazardous substances emitted by machinery — Part 2: Methodology leading to verification procedures

ISO 14159:2002 Safety of machinery — Hygiene requirements for the design of machinery

ISO/TS 14415:2005 Ergonomics of the thermal environment — Application of International Standards to people with special requirements

ISO 14505-2:2006 Ergonomics of the thermal environment — Evaluation of thermal environments in vehicles — Part 2: Determination of equivalent temperature

ISO 14505-3:2006 Ergonomics of the thermal environment — Evaluation of thermal environments in vehicles — Part 3: Evaluation of thermal comfort using human subjects

ISO 14738:2002 Safety of Machinery – Anthropometric requirements for the design of workstations at machinery

ISO/IEC 14754:1999 Information technology — Pen-Based Interfaces — Common gestures for Text Editing with Pen-Based Systems

ISO/IEC 14755:1997 Information technology — Input methods to enter characters from the repertoire of ISO/IEC 10646 with a keyboard or other input device

ISO 14835-1:2005 Mechanical vibration and shock — Cold provocation tests for the assessment of peripheral vascular function — Part 1: Measurement and evaluation of finger skin temperature

ISO 14835-2:2005 Mechanical vibration and shock — Cold provocation tests for the assessment of peripheral vascular function — Part 2: Measurement and evaluation of finger systolic blood pressure

ISO 14915-1:2002 Software ergonomics for multimedia user interfaces – Part 1: Design principles and framework

ISO 14915-2:2003 Software ergonomics for multimedia user interfaces – Part 2: Multimedia navigation and control

ISO 14915-3:2002 Software ergonomics for multimedia user interfaces – Part 3: Media selection and combination

ISO 15265:2004 Ergonomics of the thermal environment — Risk assessment strategy for the prevention of stress or discomfort in thermal working conditions

ISO/IEC 15411:1999 Information technology — Segmented keyboard layouts

ISO/IEC 15412:1999 Information technology — Portable computer keyboard layouts

ISO/IEC TR 15440:2005 Information technology — Future keyboards and other associated input devices and related entry methods

ISO 15534-1:2000 Ergonomic design for the safety of machinery — Part 1: Principles for determining the dimensions required for openings for whole-body access into machinery

ISO 15534-2:2000 Ergonomic design for the safety of machinery — Part 2: Principles for determining the dimensions required for access openings

ISO 15534-3:2000 Ergonomic design for the safety of machinery — Part 3: Anthropometric data

ISO 15535:2006 General requirement for establishing anthropometric databases

ISO 15536-1:2005 Ergonomics — Computer manikins and body templates — Part 1: General requirements

ISO 15537:2004 Principles for selecting and using test persons for testing anthropometric aspects of industrial products and designs

ISO 15664:2001 Acoustics — Noise control design procedures for open plant

ISO/TS 15666:2003 Acoustics — Assessment of noise annoyance by means of social and socio-acoustic surveys

ISO 15667:2000 Acoustics — Guidelines for noise control by enclosures and cabins

ISO/IEC 15897:1999 Information technology — Procedures for registration of cultural elements

ISO/TS 15694:2004 Mechanical vibration and shock — Measurement and evaluation of single shocks transmitted from handheld and hand-guided machines to the hand-arm system

ISO/TS 16071:2003 Ergonomics of human-system interaction – Guidance on accessibility for human-computer interfaces

ISO/TR 16982:2002 Ergonomics of human-system interaction — Usability methods supporting human-centered design

ISO 17624:2004 Acoustics — Guidelines for noise control in offices and work-rooms by means of acoustical screens

ISO/IEC 18021:2002 Information technology — User interfaces for mobile tools for management of database communications in a client-server model

ISO/IEC 18035:2003 Information technology — Icon symbols and functions for controlling multimedia software applications

ISO/IEC 18036:2003 Information technology — Icon symbols and functions for World Wide Web browser toolbars

ISO/PAS 18152:2003 Ergonomics of human-system interaction — Specification for the process assessment of human-system issues

ISO/TR 18529:2000 Ergonomics — Ergonomics of human-system interaction — Human-centered lifecycle process descriptions

ISO/TR 18569:2004 Safety of machinery — Guidelines for the understanding and use of safety of machinery standards

ISO 19353:2005 Safety of machinery — Fire prevention and protection

ISO/TR 19358:2002 Ergonomics – Construction and application tests for speech technology

ISO/IEC TR 19764:2005 Information technology — Guidelines, methodology and reference criteria for cultural and linguistic adaptability in information technology products

ISO 20282-1:2006 Ease of operation of everyday products — Part 1: Design requirements for context of use and user characteristics

ISO 20282-2:2006 Ease of operation of everyday products — Part 2: Test method for walk-up-and-use products

ISO/TS 20646-1:2004 Ergonomic procedures for the improvement of local muscular workloads — Part 1: Guidelines for reducing local muscular workloads (available in English only)

ISO 20685:2005 3-D scanning methodologies for internationally compatible anthropometric databases

ISO 21469:2006 Safety of machinery — Lubricants with incidental product contact — Hygiene requirements

ISO/IEC 24738:2006 Information technology — Icon symbols and functions for multimedia link attributes

Index